MOBILITY 2040

EXPLORING THE EMERGING TRENDS RADICALLY TRANSFORMING TRANSPORTATION SYSTEMS IN THE US

GALO BOWEN

NEW DEGREE PRESS

COPYRIGHT © 2021 GALO BOWEN

All rights reserved.

MOBILITY 2040

Exploring the Emerging Trends Radically Transforming Transportation Systems in the US

ISBN   978-1-63676-939-4   *Paperback*

978-1-63730-005-3   *Kindle Ebook*

978-1-63730-107-4   *Ebook*

*To my parents, for teaching me the meaning of selflessness,
and raising me to believe anything is possible.*

CONTENTS

———

INTRODUCTION

———

Commuting is the nightmare that interrupts the American Dream.

My story begins after my move from my home country of Ecuador to Northern Virginia in August 2007 to pursue my college education. I needed to rely on public transportation to travel to and from my classes in Vienna (VA). What would normally be a fifteen-minute drive over ten miles by car became a four-hour round trip commute each day. Running the numbers, my time spent on buses and bus stops equaled that of a part-time job, or twenty hours per week, and over eighty hours per month. As such, it did not take me long to start perceiving Virginia bus stops and public buses as pseudo-stationary public libraries as I would do most of my school reading and assignments during my commute.

For many people in the United States like myself, it's no secret the nation's transportation system is dysfunctional. A high vehicle dependency across US metropolitan areas and an inefficient public transit system have resulted in an exhaustive list of challenges the transportation sector faces, including traffic congestion, transportation inaccessibility, and devastating environmental effects. Unfortunately, over the past decades, limited effort has been directed toward creating tangible, systemic improvements to slow the decline of America's transportation systems.

In spite of this, I have no intentions of portraying myself as a *helpless victim* of the US transport sector. After all, I'm sure I'm only one of many who have experienced similar and even much worse commuting journeys, all while noticing the poor and scary condition of a large part of our transit infrastructure.

Quite the contrary, I see so many promising opportunities arising from new and cutting-edge technologies. I'm a leading evangelist for the exciting future of transportation ahead of us.

Hybrid vehicles, electric vehicles, followed by the internet of things (IoT), big data, and artificial intelligence (AI), among other emerging technologies, have been gradually penetrating the transportation space within the last decade. With the explosion of data from mobile devices, software companies are being able to leverage that data to generate key insights across transportation planning and traffic engineering to create a positive impact in the transport ecosystem.

There are so many up-and-coming tech companies right now that are energizing this space: StreetLight Data in San Francisco, INRIX in Washington, Uber, Tesla, and many others are all pushing groundbreaking changes. This provides hope for continued development of the entire transportation system.

Yet the path toward an efficient, sustainable, and inclusive transportation system will not be as easy a task, as it presents many roadblocks. In this book, we'll explore the exciting technological trends energizing this sector, in addition to the depths of the problems with current transportation systems.

IMMOBILITY 2020: UNSUSTAINABLE, INEFFICIENT, EXPENSIVE AND EXCLUSIVE

Speaking to the negative environmental impacts of transportation, the largest share of greenhouse gas (GHG) emissions in the US belongs to the transport sector.[1] According to the United States Environmental Protection Agency (EPA), over the past (close to) three decades GHG emissions from transportation saw the largest increase in absolute terms than any other sector.[2]

Traffic congestion, a more perceptible phenomenon, is another pressing transport-related challenge in US roads, resulting in significant time losses for commuters which, in turn, translate into considerable economic damage for them.

---

1   "Sources of Greenhouse Emissions," United States Environmental Protection Agency (EPA), last modified December 4, 2020.

2   "Transportation and Climate Change," Carbon Pollution from Transportation, United States Environmental Protection Agency (EPA), last modified November 20, 2020.

In less encouraging news, the crumbling US infrastructure and public transit system reflect a substantial $90 billion backlog in maintenance.[3]

As if the transport landscape wasn't complicated enough, close to half of the US population still has no access to public transportation.[4]

Carbon emissions, traffic congestion, precarious infrastructure, and transportation inaccessibility represent a few of the key challenges the transportation ecosystem faces. The disclosed facts are undoubtedly discouraging. No wonder everyone thinks the transportation system in the US is too broken and too large to be fixed.

TECHNOLOGY IS CHANGING THE GAME

While I do agree the US transportation system is limping along, I also believe innovative companies are starting to make slow and steady progress on solutions, which suggests we can dig ourselves out of the current predicament. I deposit my trust in those ingenious ideas and wholeheartedly believe with further investment and regulatory support they can help provide sustainable transportation solutions for generations to come.

---

3   Sean Slone, "Top 5 Issues for 2018: Transportation & Infrastructure: The Precarious Condition of US Infrastructure," *Sean Slone's blog, The Council of State Governments*, January 21, 2018.

4   "Public Transportation Facts," American Public Transportation Association (APTA), accessed September 12, 2020.

Consider Uber Technologies, Inc., an American ride-hailing company, which has recognized existing gaps in the American transportation industry and has put tremendous work into helping bridge these gaps. The company offers services including peer-to-peer ridesharing, food delivery, and a micromobility system with electric bikes and scooters. While Uber today is a multinational company with a reported revenue of over $10 billion in 2020, the firm started back in 2010 with a basic transportation concept of connecting drivers and passengers through a mobile app platform.[5]

Likewise, chances are most of you have heard of Tesla, Inc. Or, maybe the name of Elon Musk—the company's early investor and CEO—resonates even more. Tesla is an American electric vehicle and clean energy company which specializes in electric vehicle (EV) manufacturing and battery energy storage from home to grid scale. In the past few years, a lot of attention has been centered in Tesla Autopilot, a suite of advanced driver-assistance system features with the ultimate goal of bringing self-driving, autonomous vehicles (AVs) to reality. Similarly, Nuro, an American robotics company, has embarked on the mission of accelerating the benefits of robotics for everyday life, and their first step is a self-driving vehicle designed for local goods transportation.

Transit agencies are implementing microtransit solutions to improve the riders experience by operating small-scale, on-demand public transit services that can offer fixed routes

---

5   "Uber Announces Results for Fourth Quarter and Full Year 2020," Uber Investor, Uber Technologies, Inc., press release, February 10, 2021, on the Uber Technologies, Inc. website, accessed February 12, 2021.

and schedules, as well as flexible routes and on-demand scheduling. For instance, the GO!Bus Plus six-month pilot program was launched in July 2020 in Grand Rapids, Michigan. This initiative allowed riders to reserve trips on-demand through a single mobile app connected to various transportation providers. The program will streamline the process of scheduling medical transportation. It is a partnership between key stakeholders including: The Rapid, a public bus company; Kaizen Health, a healthcare logistics platform; the Disability Advocates of Kent County, and the city of Grand Rapids.[6,7]

In short, we are gradually witnessing the potential of technology to disrupt and/or sustain the transportation industry. The concepts of shared mobility, electric vehicles, autonomous driving, and microtransit are being explored thoroughly by technology and mobility experts in an effort to revitalize the transportation sector and address core mobility gaps in the US.

OPPORTUNITIES FOR ADVANCEMENT

There are numerous reasons and personal experiences that have led me to be compelled to write on such a complicated and thought-provoking topic like transportation in the United States.

---

6   "Pilot Program to Boost Mobility Options for People with Disabilities," Mibiz, July 13, 2020.

7   "New App Offers The Rapids GO!Bus Passengers Convenience and Less Wait Time," The Rapid. press release, August 5, 2019, on The Rapid website, accessed September 8, 2020.

One experience in particular that triggered my interest in exploring ways to improve and empower the US transportation system is tied to my global residency project at the Georgetown McDonough School of Business. During my time there, I was a consultant to Hinduja Group, an Indian conglomerate company, and Ashok Leyland, its subsidiary in the automotive sector and manufacturer of electric buses.

My team and I identified an opportunity to the emerging trend of ride-hailing, on-demand shared mobility and explored an opportunity in Mobility as a Service (MaaS), also referred to as integrated multi-modal mobility. I must admit I initially questioned the company's plan to enter a new transportation space. Why would a firm consider entering a sector outside its core competencies? How could a "simple" bus manufacturer penetrate the ride-hailing arena? It didn't take me long to realize that, indeed, the emergence of data coming from mobile devices and the momentum gained by on-demand shared mobility and MaaS represented an opportunity for our client. The Hinduja Group was being given a chance to consider reshaping its role as more than an original equipment manufacturer (OEM) and electric bus maker.

The conclusion I reached is tech-driven (e.g., mobile app, e-ticketing), consumer-centric mid- and long-distance bus services and companies had been gradually expanding in India. Our financial model forecasted the viability and long-term profitability of the "electric ride-hailing" business for our client. By leveraging Ashok Leyland's manufacturing capabilities and synergies with Hinduja Tech, its subsidiary in digital technologies provider for the automotive space, the Indian conglomerate would be able to vertically integrate its supply chain. Hinduja Group could, in fact, supply electric buses and launch a mobile app, both key for the on-demand mobility ecosystem, and thus establish itself as the first entrant in the bus segment space.

Another intriguing example is an app called Whim. Residents of Helsinki (Finland) can use Whim to plan and pay for all modes of public and private transportation within the city including train, taxi, bus, carshare, and bikeshare. Whim was developed by MaaS Global, commonly held as the world's first true Mobility as a Service operator. Anyone with the app can enter a destination and select his or her preferred mode of transportation. In cases where no single transport mode covers the door-to-door journey, a combination can be used. Users can either pre-pay for the service as part of a monthly mobility subscription or pay as they go using a payment account linked to the service.

A transportation behavior study conducted by MaaS Global caught my attention. The report indicated while 48 percent of all trips by Helsinki metropolitan area residents were made by public transportation, Whim users rode public transportation more than their counterparts at 63 percent.[8] To me, this was a clear indicator of the positive impact of Mobility as a Service on public transportation ridership. As such, it drove me to more deeply explore this emerging mobility trend where different modes of transportation are centralized through a single consumer endpoint.

"That sounds great, but we are not Finland!"

---

8   Ari Hartikainen et al., "Whimpact," *Ramboll*, May 21, 2019.

I can hear you exclaiming, but I've got news for you. It turns out given the perceived opportunity in shared mobility and MaaS, automakers in the US have already multiplied their efforts as a way to protect themselves should there be a substantial shift toward multimodal transportation and away from private car ownership. Original equipment manufacturers (OEMs) are already experimenting with several business models. Vehicle manufacturers like Volvo, Nissan, General Motors (GM), and Honda have started investing in, partnering with, and acquiring mobility and tech companies (e.g., Uber, Cruise, Waymo), as well as creating mobility subsidiaries.[9] In short, as they have recognized gaps and opportunities in the transport sector, automakers have started reshaping their roles as more than vehicle manufacturing companies.

MOBILITY 2040: TOGETHER WE CAN GET THERE

A significant part of my interest and curiosity in mobility in the United States lies on my recent experience of closely witnessing the intersection of emerging technologies and transportation. I am strongly convinced by effectively leveraging the technology, and with the mobility stakeholders already in place, the fragmented US transportation system can be revitalized to drive mobility for all the population over the next few decades.

Fundamentally, there are specific mobility trends I believe will help us empower transportation in the United States. Electric vehicles (EVs), autonomous vehicles (AVs), micromobility,

---

9   "Disrupted by Mobility Startups, Automakers Reshape Their Roles," Center for Automotive Research (CAR), May 4, 2018.

microtransit, shared mobility, and, specifically, Mobility as a Service (MaaS) are all called to build together a powerful amalgamation of transport trends on the road to Mobility 2040—seamless, integrated, and sustainable mobility for all.

Yet, the future of an enhanced and fixed transportation system in the US will hardly be possible without the collaboration of many key stakeholders, including vehicle manufacturers, private mobility companies, public transportation officials, and policy makers.

Let this be a rallying call for everyone who shares the mission of transforming the US transportation system and addressing its current environmental and social impacts. By working collectively and toward the same direction, I am confident all above-mentioned stakeholders will make this "transport utopia" a long-term, fascinating reality.

This book is for the investor looking for the first-movers advantage in this dynamic space. It's for policy makers whose decisions will shape the path we take for years to come. It's for private mobility companies and startups and original equipment manufacturers who make development possible. It's for American commuters whose hearts we must win to realize a better future for our country.

PART 1:

HOW WE GOT HERE

CHAPTER 1:

THE MODERN HISTORY OF THE US TRANSPORTATION SYSTEM

"You must always be able to predict what's next and then have the flexibility to evolve."[10]

—MARC BENIOFF, SALESFORCE CEO.

While there is no United Nations (UN) official ranking for developed countries, the United States has the world's largest economy by nominal GDP and net wealth. It is the leader in rapidly evolving industries and technological advances such as in computers, pharmaceuticals, medical, aerospace, and military equipment. These factors, without a doubt, put the US among one of the most developed nations on Earth.

---

10   Marc Benioff, quoted in Scott D. Harris, "2009 Q&A: Marc Benioff, CEO of Salesforce.com," *The Mercury News*, October 23, 2009.

The US economy is also heavily dependent on road transport for moving people and goods. Personal transportation is dominated by automobiles, which operate on a network of 4.2 million miles (6.8 million km) of public roads, including one of the world's longest highway systems at 67,300 miles (108,000 km).[11] The US has the world's second-largest automobile market after China; and with 838 per 1,000 Americans, it has the third-highest rate of per capita vehicle ownership in the world only after San Marino and Monaco.[12,13]

Amidst endless discussions on the developed and yet very imperfect US transportation system, and after having lived for nearly fourteen years in the United States and having been fully dependent on public transportation at times, I cannot help but ask myself – why does a very developed nation like the US have relatively low levels of public transit ridership compared to wealthy European and Asian countries? To me this was a paradox indeed, and we'll take a closer look at what caused the US transportation system to reach this point and the factors driving such a large dependency on personal vehicles.

---

11 "Highway Statistics 2017," United States Department of Transportation Federal Highway Administration, Policy and Governmental Affairs Office of Highway Policy Information, last modified November 27, 2018.

12 "State Motor-Vehicle Registrations 2018," United States Department of Transportation Federal Highway Administration, Policy and Governmental Affairs Office of Highway Policy Information, December, 2019.

13 "Motor Vehicles Per 1000 People: Countries Compared," NationMaster, last modified 2014.

A DISORGANIZED EXPANSION

In 2011, during a conference at the Carnegie Endowment for International Peace, a foreign policy think tank with centers in Washington, DC, Moscow, Beirut, Brussels, and New Delhi, keynote speaker David Burwell (†) referred to the modern history of the US transportation system, with the Interstate Highway System as the foundation, and highlighted the associated large financial investments:[14,15]

"The modern history of the transportation system is the creation of the Interstate Highway System in 1956. The idea was every major capital in the United States should be connected with at least a four-lane interstate highway. It was a twelve-year project costing $27 billion. This was in 1956, and in fact it was only declared complete in 1991."

"The total cost was $425 billion, in 1991 dollars. So, it took longer, but it's a major infrastructure improvement and it has been the foundation of America's economic development over the last fifty years."

"When the Interstate Highway System was finally declared complete in 1991, the program was made more flexible to fund not

---

14 David Burwell was the cofounder and first president of the Rails-to-Trails Conservancy, a Washington-based organization that led nationwide efforts to convert thousands of miles of unused railroad corridors to trails and parklands, and who served on the executive committee of the National Research Council's Transportation Research Board from 1992 until 1998.

15 Burwell, David, and Shin-Pei Tsay, "Transforming Transportation for the 21st Century," Carnegie Endowment for International Peace, July 14, 2011.

only highways, but a wide variety of transportation projects from transit to buses, to bus rapid transit to streetcars to bikes."

Although Burwell expressed his admiration for such a massive and bold movement in the transportation sector, he also suggested this transit system lacked direction and focus in the years following completion of the interstate.

"There was no directive, however, for a specific additional system. The result was many states continued to just build highways or felt they could build projects rather than a system. It became highly earmarked, highly project-focused, and therefore unfocused."

On a similar note, Joseph Stromberg, former writer at Vox, an American news and opinion website owned by Vox media, shared his insights on the rapid deployment of personal vehicles across different American cities.[16]

"Most of our cities and suburbs were built out after the 1950s when the car became the dominant mode of transportation. Consequently, we have sprawling, auto-centric metropolises that just can't be easily served by public transportation."

AN OVERLOOKED PUBLIC TRANSIT

Despite the common belief the development of the US post-1950s made a poor transit system inevitable, the truth is, Stromberg claims, if we look closer at transportation history

---

16   Joseph Stromberg, "The Real Reason American Public Transportation Is Such A Disaster," *Vox*, last modified August 10, 2015.

in other countries, we realize they combined suburbs with better transit. As a matter of fact, David King, an assistant professor of urban planning at Columbia University and Arizona State University in the 1950s, stated the United States, Canada, Germany, France, the UK, and Australia were all on the same trajectory of racing toward automobile dependence, and yet in the 1960s a divergence was evident.

Stromberg suggests in the 1960s many European cities put large efforts into maintaining already existing transit systems and expanded them to growing suburbs. Similarly, newer cities in Western Canada—even though they were being designed for personal vehicles—heavily invested in light rail lines and improved quality bus service, translating into higher levels of public transportation ridership than US cities of similar size and density.

Conversely in the United States, the gradual expansion of newer cities in the West and South did not come with the necessary associated investment in public transportation. In certain big cities such expansion dismantled their existing transit systems, replacing streetcar (aka trolley or tram) lines with highways to accelerate commutes from the suburbs. An example of this is Boston; transit blogger Alon Levy explains:[17]

"In 1912, Boston had this great public transit system with four subway lines and streetcars that fed it... Then they spent the next sixty or seventy years destroying it."

---

17   Ibid.

If you ask me, the results of a long-overlooked US public transportation system are visible today. Public transit is indeed much more reliable and effective in European countries than it is in the United States, which largely explains the higher car dependency in US cities.

During my trip through Western Europe in 2012, the stark contrast for convenience and affordability of the Eurail train was apparent. It was very easy for my friends and I to mobilize inside the metropolitan areas of Paris, Madrid, Barcelona, Brussels, Berlin, and Milan by solely relying on public transit. Other than walking to places for short distances, bus and rail were the way to go. The idea of renting a car was not even an option we would consider. Indeed, the robust European mass transit system represented, without a doubt, the best transport alternative.

Could I say the same about US metropolitan regions? Not precisely. Apart from certain exceptions including New York City's subway and, to a lesser extent, Chicago's "L" and Washington, DC's metro system, a rather inefficient and dysfunctional public transit system has been the norm. When traveling to cities like Miami, Houston, and Orlando, it has become quite evident to me the necessity to secure a car rental if I wanted to get around easily and quickly.

Interestingly enough, these last three places are not even among the ten least car-free large metros. Birmingham (AL), Nashville (TN) and Raleigh (NC) appear as the large metropolitan areas where it is the hardest to go without a car, with Metro Car-Free Indexes of 0.205, 0.274 and 0.283

respectively.[18] The Metro Car-Free Index measures the US cities where people go without a car in the largest numbers. For reference, New York City's is 0.923 with 1.0 being the highest.

On a further note, Stromberg claims US cities perceive public transportation as welfare, as a government aid to their citizens, which explains the decline of transportation services. According to Stromberg, the difference between the United States and Europe started in the 1950s when streetcar and bus companies went bankrupt and were taken over by municipalities in many US cities. The decline of transit services can be attributed to a few distinct reasons, Stromberg continues. Companies were not allowed to raise their fares and were responsible for roads maintenance, while personal car traffic rapidly increased making streetcars slow.

"Once just 10 percent or so of people were driving, the tracks were so crowded that [the streetcars] weren't making their schedules."[19]

—PETER NORTON, TRANSPORTATION HISTORIAN

Norton's statement is quite the revelation. With only 10 percent of people driving a personal vehicle, streetcars did not appear to be a very reliable transportation method. This would certainly move more people to cars, pushing public transit further behind schedule, and thus contributing to the disappearance of streetcars altogether.

---

18 Richard Florida, "The Best and Worst US Places to Live Car-Free," Bloomberg CityLab, *Bloomberg*, September 24, 2019.

19 Peter Norton, quoted in Joseph Stromberg, "The Real Story Behind The Demise of Americas Once-Mighty Streetcars," *Vox*, May 7, 2015.

According to Stromberg, cities took over these companies, converted streetcar lines into buses, and considered these systems as part of a welfare service for people who could not afford to drive. This perception has not changed, with the exception of a few cities like New York City and Washington, DC. Indeed, many politicians, rather than considering public transit as a necessity for a developed society, see it as a government aid program aimed to help the poor who cannot afford a car.

The consequence of this mentality? American cities heavily subsidize public transit—no more than 30 percent to 40 percent of operating costs are covered by fares. This has a negative effect since it prevents local agencies from charging fares high enough to provide efficient and effective transit service to all individuals, not just those who cannot afford a personal car to drive.

In David King's words:

"Transit in the US is caught in a vicious cycle. We push for low fares for social reasons, but that starves the transit agency, which leads to reduced service."

Stromberg considers this is the same dilemma faced by the streetcar companies back in the 1950s and explains why numerous US cities bus and rail systems have limited operating hours and frequency, even those having relatively extensive networks and several stops. Having a bus coming every thirty minutes is, in fact, considered acceptable for people who cannot afford anything else.

In contrast, London and Toronto have transit systems with higher fares and a more frequent service, which results in a more attractive option for people who own cars. In the case of Paris, each municipality is legally obliged to pay the transit agency the difference between its fares and operating costs, helping it to provide an efficient service while keeping fares down. Other cities, like Seattle, have experimented with charging lower fares to people with reduced income.

GOVERNMENT SUBSIDIES, VEHICLES TAXES, AND TECHNOLOGICAL FOCUS

Likewise, according to Ralph Buehler, associate professor and chair of Urban Affairs and Planning at Virginia Tech's School of Public and International Affairs (SPIA), for nearly fifty years government subsidies for driving have lowered its cost and increased demand in the United States.[20]

"...gas taxes, tolls, and registration fees have covered only about 60 or 70 percent of roadway expenditures across all levels of US government. The remainder has been paid using property, income, and other taxes not related to transportation."

On the other hand, in European countries, Buehler suggests drivers typically pay more in taxes and fees than governments spend on roadways.[21]

---

20 Ralph Buehler, "9 Reasons the US Ended Up So Much More Car-Dependent Than Europe," Bloomberg CityLab, *Bloomberg*, February 4, 2014.

21 Eric Jaffe, "These 2 Charts Prove American Drivers Don't Pay Enough for Roads," Bloomberg CityLab, *Bloomberg*, September 18, 2013.

Furthermore, Stromberg suggests the US political system is biased against public transit and describes the difference with Canada and Europe.

"There are other quirks of American politics that have arguably led us to under-invest in transit. Because it's often seen as welfare, investing in mass transit has become a politically charged issue, with conservatives unwilling to spend on what they see as a social program for the urban poor."

"This doesn't really happen in other countries, at least not to the same extent. While there is some debate over transit spending in Canada and Europe, politicians on the right are much less hostile to the idea. It's much more of a bipartisan cause like, say, road building in the US."

The federal government also plays an important role in driving transportation policy, says Stromberg. Federal policy is often largely aimed to meet rural interests rather than urban priorities.

"The postwar directive to demolish urban neighborhoods to build highways came from the Department of Commerce, not from individual cities, and has been carried out by the Department of Transportation. By contrast, in Canada there is no corresponding national department, and regional bodies have greater say in transportation planning."

Along the lines of the role of governmental policy in transportation, Ralph Buehler explains how rather than changing societal behavior, policy in the US has focused on technological changes to alleviate the problems associated with vehicle travel.

"For example, [in the US] responses to air pollution or traffic safety consisted of technological fixes—such as catalytic converters, reformulated cleaner fuels, seat belts, and air bags—that let people keep driving as usual."

In contrast, besides adopting the same technological requirements as their American counterparts, European countries implemented policies that encouraged behavioral shifts. For example, Buehler suggests, Europe designated car free zones and bike networks, significantly lowered speed limits in entire neighborhoods, and reduced car parking.

Lastly, differences in vehicles taxes can also explain a higher car demand in the United States than in European countries. Taxation of car ownership and government use has, in fact, traditionally been higher in Europe and helped lower car travel demand. Likewise, consider a gallon of gasoline is more than twice as expensive in Europe than in the US—$7.32 (Norway), $7.07 (Netherlands), $6.69 (Italy), $6.46 (France), and $6.17 (UK) versus $2.91 in the United States, in the fourth quarter of 2019.[22] In addition, Buehler claims in Europe gas tax revenue typically contributes to the general fund, meaning roadway expenditures compete with other government expenditures. On the other hand:

---

22 "Gasoline Prices In Selected Countries Worldwide In 4th Quarter of 2019," Statista, accessed September 10, 2020.

"In many US states and at the federal level, large parts of the gas tax revenue are earmarked for roadway construction, assuring a steady flow of non-competitive funds for roads."

KEY TAKEAWAYS

In summary, the modern history of the United States transportation system has taught us that:

- In light of its rapid development through the creation of the Interstate Highway System, the subsequent suburban expansion lacked direction and focus.
- This translated into a high vehicle dependency and decreased levels of public transit ridership across US metropolitan areas—relatively lower than in European and Canadian cities.
- Vehicle taxes, government subsidies, a technological exclusive policy focus, and an overall lack of investment in public transit all further contributed to the US automobile dependency and disregarded public transit.

From there, we will go over the implications of a high per capita vehicle ownership rate in the United States and discuss the consequences of an overlooked public transportation ecosystem. We will explore the arising transport-related problems that have led people to consider the US transportation system as completely broken and impossible to fix.

CHAPTER 2

PROBLEMS WITH THE SYSTEM TODAY

"If we're going to talk about transport, I would say that the great city is not the one that has highways, but one where a child on a tricycle or bicycle can go safely everywhere."[23]

—*ENRIQUE PEÑALOSA*, URBANIST AND
FORMER MAYOR OF BOGOTÁ

The US Interstate Highway System is a highly developed network of control-access highways. While comparisons are odious, it has always been inevitable for me to contrast such a massive infrastructure system to what is in place in my home country of Ecuador and the rest of the developing world. I have, in fact, constantly wished for Ecuadorian authorities to implement a similar highway system to the one in the United States.

---

23 Enrique Peñalosa, quoted in Matthew Roth, "Enrique Peñalosa Urges SF to Embrace Pedestrians and Public Space," *StreetsBlog SF*, July 8, 2009.

As explained in an earlier chapter, the US transportation system has, nevertheless, become a victim of its own success. The hasty and unfocused highway expansion, and the associated surge in vehicle dependency and overlooked public transit, have led the US population to face a myriad of transport-related challenges, many of which I have experienced for fourteen years living the American Dream.

TRAFFIC CONGESTION

Chances are that you will easily relate to the issue of traffic congestion. Just think for a moment how many times and for how long you get stuck in traffic going to the office, driving back to your home, riding on a vacation road trip with your family and friends, picking up your groceries, or running your weekly errands.

I know for a fact I personally have spent more time in traffic jams than I would have liked. I experienced rush hour commutes from my former office at Owings Mills in a Maryland suburb to home in downtown Baltimore, constantly drove through downtown Washington, DC and, on many occasions, New York City and Miami. Likewise, numerous road trips to Maryland and Florida beaches where traffic jams were the norm, are just a few examples of my experience dealing with traffic congestion in the US.

Similarly, but worse, I remember my roommate Gregory complaining during our time in Baltimore about his commute back to our apartment from his office in the neighborhood of Fells Point. To avoid the unpleasant situation of getting stuck in traffic, he would opt to give up his car and go on a one-mile bike ride to the water taxi station where he would jump on a boat. Gregory would then cross the waters of the Patapsco River and finish his commute to his workplace with a short five-minute walk.

"My office is only three-miles away [from his apartment in Locust Point, Baltimore] and yet it takes me at least one hour to get home during rush hour. I don't mind biking to the water taxi to save myself at least thirty minutes of commute time."

Biking, water taxi, and walking—that sounded like a whole commuter adventure to me. But who knew multi-modal transportation could be a much better alternative than relying solely on a personal car? Gregory did. He figured it out only a few days upon starting his new job, and I even remember a few occasions when I had to go pick him up in my car because he had missed the last running water taxi. That was a long drive for me as well, but hey, the Ubers and Lyfts were not so popular back in 2013 and I had to give my friend a hand.

As Gregory started gradually having more responsibilities at work and hence his work hours extended, making it impossible for him to catch the last water taxi scheduled at 6:00 p.m., he had no choice but to start driving his Ford Focus again. Interestingly enough, his commutes back home became a lot shorter than before as he was working longer hours, leaving the office past 6:00 p.m., and avoiding peak times. This is certainly the case of many people who stay away from heavy traffic jams by coordinating their work hours with peak driving hours.

Moreover, parking difficulties and traffic congestion are interrelated. When returning home from work during rush hour, there were times I found myself driving around looking for a street parking space for up to forty-five minutes in the neighborhood of Federal Hill (Baltimore). This created additional delays and impaired local circulation.

Gregory and I have evidently witnessed first-hand the heavy traffic congestion in US cities. Truth is, however, we are only two of dozens of millions of people who have to constantly deal with this phenomenon which repeats across numerous metropolitan areas in the United States and is, in fact, much worse in other cities.

According to the 2019 Traffic Scorecard report by INRIX—a leader in connected car services and transportation analytics—Boston and Chicago are among the top ten most gridlocked cities in the world. At the US level, drivers in Boston, Chicago, and Philadelphia lost the most time in 2019 battling traffic congestion during peak commute periods compared to free-flow conditions with 149, 145, and 142 hours, respectively.[24]

Similarly, Philadelphia, Washington, DC, and San Francisco all had the slowest last mile travel speeds in the country (10 MPH), defined as the speed at which a driver can expect to travel one mile into the central business district during peak hours.

---

24 Trevor Reed, "INRIX Global Traffic Scorecard," INRIX, INRIX Research, March, 2020.

Furthermore, the report indicates overall, in 2019, the average American driver lost ninety-nine hours a year battling traffic, during peak travel times of 6:00 a.m. to 9:00 a.m. and 3:00 p.m. to 6:00 p.m., costing $1,377. Nationally, drivers lost more than $88 billion in time to congestion. The estimated totals tally direct costs, such as the value of time spent in traffic and payments for extra fuel; plus indirect costs, including higher delivery spending for goods and services that companies pass along to customers.

INRIX's report lastly points out Los Angeles has three of the top ten most congested corridors in the country, followed by New York City and Chicago with two each.

Based on the overall findings, the US ranked as the most traffic-congested developed nation in the world. After all, Gregory and I were not overreacting to traffic, were we?

Further, have I mentioned the associated vehicular accidents? In 2019, crashes involving motor vehicles killed 36,096 people and injured another 2.74 million in the US.[25]

Likewise, a study conducted by the National Highway Traffic Safety Administration found in 2010 alone, fatalities, injuries, and property damage from motor vehicle accidents cost the economy $242 billion.[26] This is equivalent to $784 for every

---

25 *Overview of Motor Vehicle Crashes in 2019*, National Center for Statistics and Analysis, Traffic Safety Facts Research Note, Report No. DOT HS 813 060 (Washington DC: National Highway Traffic Safety Administration, 2020).

26 Lawrence Blincoe et al., *The Economic and Societal Impact of Motor Vehicle Crashes, 2010*, Report No. DOT HS 812 013 (Washington, DC: National Highway Traffic Safety Administration, 2015).

person living in the United States and 1.6 percent of the US GDP for 2010, or the total economic output of the nation. Considering quality of life valuations, the total value of societal harm from motor vehicle crashes was $836 billion in 2010, 20 percent higher than the estimated costs in 2000.[27,28]

In short, traffic congestion, long commutes, parking difficulties, and automobile accidents are challenges the US population faces in their day-to-day lives.

AIR POLLUTION

Breathing is the most universal experience there is and most of the time we don't even think about it. In too many places it carries a hidden danger. From heart attacks and strokes to dementia and premature birth, scientists have linked air pollution to a long and growing list of health problems. Air pollution has, in fact, become an obstacle to the quality of life and the health of urban populations.

In her 2019 Tedx Talk in London, environmental journalist Beth Gardiner explains the air we breathe is killing us.[29]

"I would have been pretty ready to believe, I think most of us would be, that dirty air could trigger asthma attacks and other breathing problems to maybe even lung cancer, but what shocked me was how much further the effects actually go."

---

27 Ibid.

28 "Motor Vehicle Crashes Cost the US Nearly $1 Trillion/Year," Miller Kory Rowe LLP, April 27, 2017.

29 *TEDx Talks,* "The Air We Breathe Is Killing Us—But It Doesn't Have To | Beth Gardiner | TEDxLondon," June 11, 2019, video, 12:28.

"The evidence is overwhelming. Scientists have linked air pollution to increased rates of heart attacks, strokes, many kinds of cancer, dementia, Parkinson's disease, miscarriages, premature birth, and much more."

I quickly resonated with Gardiner's discussion on air pollution—not because one of my loved ones or I was diagnosed with a disease linked to air pollution, thankfully.

In the introductory chapter of this book, I mentioned I had the opportunity to visit India for two weeks in early 2020. While I have to admit I had given up on jogging then, I could overhear a number of my friends having a recurring conversation that going for a run in downtown Mumbai was out of the question.

"Are you really going to go run in this weather?"

"Have you even checked the air quality index [AQI]?"

An air quality index (AQI) is used by government agencies to communicate to the public how polluted the air currently is or how polluted it is forecast to become. Public health risks increase as the AQI rises. This is the case of the United States Environmental Protection Agency (EPA) that uses the AQI to report air quality, which is divided into six categories indicating increasing levels of health concern. An AQI value below fifty means the air quality is good.[30]

---

30 "AQI Air Quality Index," United States Environmental Protection Agency (EPA), Office of Air Quality Planning and Standards, EPA-456/F-14-002, February, 2014.

With an AQI of over 130, the air of Mumbai fell in the category of "unhealthy for sensitive groups" and, hence, exercising outdoors was clearly not a smart choice. My peers and I could tell and feel the difference between India's metropolis air and the Western world's relatively cleaner environment.

Similarly, in the summer of 2019 I visited Mexico City as part of a career trek sponsored by Georgetown's Latin-American Business Association (LABA).

Discussions around going on a quick run around Mexico's capital quickly vanished. Contrary to Mumbai, the thickness of the air in Mexico City was easily notable to the twenty-eight pair of eyes composing our group. For those of us who were visiting the city for the first time, the giant smog clouds in the sky were assumed to be caused by massive fires when, in reality, they had *always* been there as a clear proof of the exorbitant pollution levels in the capital of the Mexican people.

"This is America and air pollution is not that big of a concern here!"

It's a repeated phrase I've gotten to hear from many people inside the United States. To be fair, I have not experienced in the US a similar situation to India or Mexico where someone has warned me not to run outdoors to avoid risking my healthy lungs to polluted air.

Although the US appears far from the top of the world's most polluted countries list, it ranked eighty-seventh according to IQ*Air* aggregated data with 9.04 μg/m^3.[31] Truth is people in the US are not immune to air pollution's negative effects like Gardiner explains.

"Even in the US, where the air is relatively clean, MRIs on women who breath moderate levels of pollution found the areas of their brains, known as the white matter, were as small as they'd be from a year or two of additional aging."

Similarly, as of 2019 more than one hundred forty-one million Americans lived in places with unhealthy levels of air pollution according to the American Lung Association's 2019 State of the Air report.[32] This is an increase of more than 7.2 million people from the number reported in 2018. As a matter of fact, I think of my California friends and how with the (now annual) seasonal wildfires in the Golden State and the US Pacific Northwest, there is unfortunately a time of the year when air pollution levels skyrocket. In addition, the central valley in California where most of the nation's produce is grown is one of the most air polluted places in the nation and often has AQI ratings of hazardous levels, even when there aren't wildfires happening.

Even though the United States is not among the most polluted countries on Earth, it accounts for 15 percent of global carbon dioxide (CO_2) emissions (5.41 GT) according to data

---

31   "World's most polluted countries 2019 (PM2.5)," IQAir, accessed September 25, 2020.

32   "More than 4 in 10 Americans Live with Unhealthy Air; Eight Cities Suffered Most Polluted Air Ever Recorded," American Lung Association. press release, April 24, 2019, on the American Lung Association website, accessed August 25, 2020.

compiled by the International Energy Agency, ranked second on the list only after China.[33] Interestingly enough, vehicles, airplanes, ships, and other forms of transport emit more greenhouse gases than any other sector of the economy in the United States, with 28.2 percent according to the United States Environmental Protection Agency.[34] That share is growing because other sectors of the economy are reducing their emissions faster than transportation.

In short, the Earth's changing climate poses one of the most important threats humanity has ever faced. To avoid catastrophic changes, all sectors of the economy, and particularly transportation, need to find ways to become more efficient so as to not further increase pollution problems as the population continues to grow.

TRANSPORTATION INACCESSIBILITY

Traffic congestion and long commutes are certainly annoying experiences to have. But after all, shouldn't those of us able to afford a car and move around at our ease consider ourselves privileged? In fact, what if I told you according to a 2019 report by the National Academy Press (NAP), roughly 20 percent of households with incomes below $25,000 lack a car?[35]

33 "Each Country's Share of CO2 Emissions," Union of Concerned Scientists, last modified August 12, 2020.

34 "Sources of Greenhouse Emissions," United States Environmental Protection Agency (EPA), last modified December 4, 2020.

35 *Critical Issues in Transportation 2019*, National Academies of Sciences, Engineering, and Medicine (Washington, DC: The National Academies Press, 2018).

Cost of car ownership includes far more than the actual price (MSRP) we pay for them. Consider the required insurance premiums, licensing and registration, and personal property taxes. This is in addition to fuel, maintenance, and repair costs (e.g., oil changes, tires), and finance charges for those who pursue the financing route. According to American Automobile Association (AAA) research, the average annual cost of vehicle ownership in the US in 2020 was $9,561, or $796.75 a month, representing the highest cost associated with new vehicle ownership since AAA began tracking expenses in 1950.[36]

As such, I disagree, partially at least, with the commonly held belief owning a car is no longer a luxury, but a necessity. More than forty million Americans live in poverty, and outside central cities an automobile is essential for access to jobs, healthcare, and other resources.[37] Owning a car should not be taken for granted.

Also speaking to the economic and financial sacrifice involved in purchasing a car, the Federal Reserve Bank of New York reported in February of 2019 that seven million Americans were ninety days or more behind on their auto loan payments.[38] Many people are, in fact, not able to own a car or are at risk of losing it due to delinquency.

---

36 Ibid.
37 Heather Long, "A Record 7 Million Americans Are 3 Months Behind On Their Car Payments, A Red Flag For The Economy," *The Washington Post*, February 12, 2019.
38 "Your Driving Costs, 2020," American Automobile Association (AAA), December 14, 2020.

In addition, and according to the same NAP study, nearly forty million Americans have some form of disability, of whom more than sixteen million are working age. The population is also aging: the number of people older than sixty-five will increase by 50 percent from forty-nine million in 2020 to seventy-three million by 2030.

Transportation accessibility also goes beyond someone's ability or inability to afford a car. As a matter of fact, the US public transportation network is not fully inclusive, and its accessibility is very disproportionate across populations. If you live in a metropolitan region like New York City, Chicago or Washington, DC like I do, chances are that using public transportation—bus and/or metro (aka subway)—is fairly easy and, on many occasions, can be a much more effective transport method than your personal Ford, Jeep, Honda, or BMW.

EARLY MORNING HUSTLES: THE TRAVELS OF CARRIE BLOUGH AND KAREN ALLEN

While investigating on US public transportation network inclusiveness, I came across a few stories I've heard many people refer to as "inspiring" but that, in my opinion, are simply depressing.

This is the case of Carrie Blough and Karen Allen, two women living sixty miles apart but both victims of transit deserts and extreme commutes.[39] Transit deserts are defined

---

39 Joseph P. Williams, "In an Unequal America, Getting to Work Can Be Hell," *The Nation*, January 29, 2019.

as areas where demand for transportation exceeds supply. They represent underserved areas of a city, not a citywide transportation shortage.

Blough is a museum curator living in Brunswick, Maryland, and starts her day at 4:00 a.m. to prepare to catch the 6:40 a.m. Maryland Area Regional Commuter (MARC) train to downtown Washington, DC. Under the best conditions with no freight rail congestion or mechanical problems, the ride is about eighty minutes one-way and it's only her first public transit misadventure.

"I think about my commute as [really starting] once I get off my train—and then I go."

After completing "round one," it takes Blough another five minutes via train or metro to arrive to the Farragut North stop, after which it's commuter hell again:

"Then I'm in motion. Then I'm fighting crowds; then I'm being stressed out."

Similarly, Allen wakes up at 5:00 a.m. to make it on time to her daily job as a housekeeper at an upscale hotel. After coffee and walking her dog, she leaves her home in Southeast DC at 6:25 a.m. to hop on TheBus—a Prince George's County, Maryland bus stopping by her neighborhood. The bus schedule, however, is very unreliable and inconsistent, leading her to give herself at least an hour to arrive to work.

"I leave my house the exact same time the [6:25 a.m.] bus leaves the station and either I meet up with the bus or I wait because

the bus is late... Sometimes its twenty minutes late, sometimes thirty. Sometimes it doesn't show up at all."

Allen claims, nonetheless, she has a plan B consisting of a different bus, NH-1, that ends its route in National Harbor. She will take it if it arrives to her stop before TheBus. Unfortunately, this is a longer route as she suggests.

"[It] goes all over the place before it hits the Gaylord...nobody likes to catch that one. It takes too long."

Thankfully for Allen, she is not required to work late hours when transit options are far more limited. TheBus is even less reliable after 7:00 p.m. and the NH-1 stops running at midnight. As such, other people do suffer the consequences of an inconsistent and unreliable public bus system.

"One lady [whose shift ended at 3:00 a.m.] used to sleep in the cafeteria until the bus starts running again."

The aforementioned commute hell stories sync well with a study conducted in 2018 by the *Smithsonian Magazine* and the Urban Information Lab at the University of Texas.[40] The report points which areas in fifty-two major US cities do not have sufficient alternatives to car ownership. Results indicate in some of the most severely affected cities, 1 in 8 residents live in what we already referred to as transit deserts.

---

40 Junfeng Jiao et al. "Dozens of US Cities Have Transit Deserts Where People Get Stranded," *Smithsonian Magazine*, March 16, 2018.

Moreover, the study found transportation deserts were present to varying degrees in all fifty-two cities considered. Forty-three percent of residents were transit dependent in transit desert block groups, on average, and surprisingly 38 percent of the population was found to be transit dependent in block groups having enough transit service to meet demand—a clear indication about the need for alternatives to individual car ownership.

For example, the analysis determined in San Francisco, 22 percent of block groups were transit deserts. This does not mean transit supply is weak within San Francisco, but it rather shows transit demand is high because many residents do not own cars or cannot drive. In certain neighborhoods, this demand is not being met. In contrast, the city of San Jose, California, has a high rate of car ownership and thus has a low rate of transit demand. The city's transit supply is relatively good, and the results indicate only 2 percent of block groups were transit deserts.

Overall, transportation inaccessibility in the US is a steady reality. From vehicle non-affordability to mainly transit deserts reinforcing inequality, the US transportation system is serving its population in a disproportionate manner.

KEY TAKEAWAYS

To summarize, the US transportation system today faces a myriad of pressing challenges including:

- Traffic congestion, long commutes, parking difficulties, and vehicular accidents.

- Air pollution—high levels of vehicular carbon emissions.
- Transportation inaccessibility—public transit and infrastructure disproportionally serving the US population.

If no actions are taken to mitigate the above transport-related issues, I'm afraid the rapid urbanization projected by the United Nations, estimating that over two-thirds of the world's population will live in cities by 2050, will just worsen the current transport situation.[41]

I do not feel discouraged whatsoever, and I would also invite you not to be. In fact, I believe the US transportation system can be revitalized, and I will explain over the next chapters how we can achieve this together.

---

41 "68% of the world population projected to live in urban areas by 2050, says UN," United Nations, May 16, 2018.

PART 2:

TRENDS TO EMPOWER MOBILITY IN 2040

CHAPTER 3:

ELECTRIC VEHICLES

"In order to have clean air in cities, you have to go electric."[42]

—ELON MUSK, TESLA CEO

We've talked about the history and current state of the US transportation system and will now look toward innovations that can help us chart a better course moving forward.

Transportation is the largest source of carbon emissions in the US According to the United States Environmental Protection Agency. In 2018, the transportation sector—including ground, air, and naval—had the largest share of greenhouse gas emissions in the country with a total of 28.2 percent.[43] Further, the Center for Climate and Energy Solution (C2ES) found passenger cars and light-duty trucks (e.g., sport vehicles, pickup trucks, and minivans) are responsible for half

---

42 Elon Musk, quoted in Lorraine Chow, "Elon Musk: You Can Easily Power All of China With Solar," *EcoWatch*, January 29, 2016.

43 "Sources of Greenhouse Emissions," United States Environmental Protection Agency (EPA), last modified December 4, 2020.

of the carbon dioxide emissions from the US transportation sector.[44]

Fortunately, the emergence of electric vehicles (EVs) has been offering a low-carbon alternative to gasoline-powered vehicles. Automakers like General Motors (GM), Ford, Tesla, Honda, and BMW are competing to reduce battery costs—which accounts for most of an EVs additional cost—increase battery range, and offer a wider range of affordable EV styles, including SUVs and minivans. Aside of the environmental benefit EVs offer, vehicle manufacturers are realizing EVs could become a better decision than traditional combustion engine vehicles from a profit perspective too.

Let's take a moment to describe the two distinguished basic types of electric vehicles (EVs) by the US Department of Energy: all-electric-vehicles (AEVs) and plug-in hybrid electric vehicles (PHEVs):[45]

> "AEVs are powered by one or more electric motors. They receive electricity by plugging into the grid and store it in batteries. They consume no petroleum-based fuel and produce no tailpipe emissions. AEVs include Battery electric vehicles (BEVs) and fuel cell electric vehicles (FCEVs)."

---

44   "Reducing Your Transportation Footprint," Center for Climate and Energy Solutions (C2ES), accessed October 5, 2020.

45   "Electric Vehicle Basics," United States Department of Energy, Office of Energy Efficiency & Renewable Energy, accessed October 5, 2020.

"PHEVs use batteries to power an electric motor, plug into the electric grid to charge, and use a petroleum-based or alternative fuel to power the internal combustion engine. Some types of PHEVs are also called extended-range electric vehicles (EREVs)."

Federal policy and subsidies have been introduced by the US government with the goal of reducing the nation's dependence in non-renewable fossil fuels and of increasing the access to clean energy technologies. In 2009, the federal government announced the availability of $2.4 billion in funding to put US vehicle manufacturers and American ingenuity to build plug-in hybrid electric vehicles (PHEVs).[46] This helped pave the way to today's all-electric vehicles. In addition, a federal tax credit incentive of up to $7,500 was introduced for individuals purchasing PHEVs. Except for models from GM and Tesla, all mainstream, mass-produced full-electric vehicles on sale in the United States qualify to this day for the full $7,500.[47] Moreover, in 2016, the federal government implemented a framework for collaboration for vehicle manufacturers, electric utilities, EV charging companies, and states all geared toward accelerating the development of EV charging infrastructure and deploying more EVs on the road.[48]

---

46 "President Obama Announces $2.4 Billion in Funding to Support Next Generation Electric Vehicles," United States Department of Energy, March, 2019.

47 John M. Vincent, "How Does the Electric Car Tax Credit Work?," US News & World Report, June 15, 2020.

48 "Fact Sheet: Obama Administration Announces Federal and Private Sector Actions to Accelerate Electric Vehicle Adoption in the United States," The White House. press release, July 21, 2016, on The White House website, accessed January 25, 2021.

As such, both the private and the public sectors are working to reduce barriers and expand EV sales.

For the remainder of this chapter, the focus will be on all-electric vehicles (AEVs). They do not depend on fossil fuels and are aimed to drastically reduce carbon-emissions in the environment. For simplicity purposes, I will refer to AEVs in the subsequent paragraphs and chapters as EVs.

BREAKTHROUGH IN EV BATTERY TECHNOLOGY

In an interview I conducted with Michael Berube, acting deputy assistant secretary for transportation in the US Department of Energy's (DOE) Office of Energy Efficiency and Renewable Energy, he shared his experience in the EV space, starting with his involvement at Chrysler in 1992 and witnessing the company's first EV deployment.

"Chrysler was cruising its first electric vehicle called the TEVan. Chrysler in the '80s had designed the minivan, this iconic vehicle, and they made an electric version. We produced maybe five hundred but it was the beginning at that time."

California had passed an electric vehicle mandate requiring automakers to have at least 2 percent of their cars electric by 1998. As such, having come out of MIT (Massachusetts Institute of Technology) a few years before, Berube says he was looking at alternatives to fossils fuels and wondering whether something could be done in the short-term future to achieve this huge increase in EVs.

"And I remember calling a professor at MIT who was best known for his work on EV batteries, 'look, is there anything on the horizon in the next four to five years that could make this possible?' Because we were coming up to this, you had to sell 2 percent of your cars electric."

"He said people had worked on battery chemistry and there wasn't anything. My sense out of that was there was just no breakthrough. Then you fast forward and that actually became true."

Before California's mandate there was not a desire or need for car manufacturers to deploy EVs, and yet the mandate pushed them to really start thinking and looking at battery chemistry. The US government and the Department of Energy started truly investigating and working on battery technology, expanding the amount of research. As such, it started with lithium-ion batteries until arriving to today's lithium-nickel-manganese-cobalt-oxide (NMC) cathode battery type, which led to the breakthrough in EV batteries. Berube explains:

"As people started working on battery technology, barriers were knocked down, and the history for the last twenty years has been a continuation of these barriers being knocked down."

Batteries are indeed the key piece for EVs deployment, and we are currently at a big turning point. As I was speaking to Berube, I remembered I had come across a General Motors (GM) announcement on the automakers future of battery technology—Ultium batteries and a completely wireless battery management system or wBMS.[49] GM is condensing the power of electronics.

In March 2020, GM gathered hundreds of employees, dealers, investors, analysts, media, and policy makers to reveal the company's new Ultium batteries which were unique in the industry.[50] Due to their large-format, pouch-style cells can be stacked vertically or horizontally inside the battery pack, allowing engineers to optimize battery energy storage and layout for each vehicle design. These batteries are flexible enough to incorporate new chemistry over time as technology changes.

wBMS aimed to improve Ultium batteries scalability and manufacturing, thus translating into significant cost savings for the company and its customers.[51] It allows individual modules in the system to communicate through a wireless network instead of traditional cables. This reduces the amount of wiring needed in battery assemblies by up to 90 percent reducing the overall vehicle weight and enabling a cleaner design and easier-to-build batteries. Further, wBMS is designed to receive new features as software becomes available in the future, something enabled via over-the-air updates.

---

49 "General Motors Future Electric Vehicles to Debut Industry's First Wireless Battery Management System," General Motors, News, September 9, 2020.
50 "GM Reveals New Ultium Batteries and a Flexible Global Platform to Rapidly Grow its EV Portfolio," General Motors, News, March 4, 2020.
51 "General Motors Future Electric Vehicles to Debut Industry's First Wireless Battery Management System," General Motors, News, September 9, 2020.

Examples like the aforementioned make Berube believe we're at the point where the batteries are getting good enough.

EV CHARGING INFRASTRUCTURE, BATTERY TRAVEL RANGE, AND CHARGING LEVELS

Chances are you have heard of the name of Elon Musk, CEO of electric vehicle and clean energy company Tesla, Inc. With a valuation of $461 billion in mid-November 2020, Tesla passed the combined market share for most of its non-EV competitors, and briefly made Elon Musk the richest man in the world.[52] Heads-up investor readers, EV technology companies can make you wealthy.

Tesla was founded in 2003 by a group of engineers who wanted to demonstrate EVs can be better, quicker, and more fun to drive than gasoline cars. During a talk in 2016, Musk discussed Tesla's business plan (aka Master Plan) he had written and published over a decade prior:[53]

"It is the only plan I thought had a chance of success, which is to start off with a low-volume car that would have a high price because we didn't have the economies of scale… Step two was to have a lower-price, higher-volume, and then step three was to make it high-volume and more affordable."

Musk successfully met his master plan's all-electric vehicles production and launching objectives and has since expanded

52 "Tesla is Now More Valuable Than 12 Top Automakers Combined," *MyBroadband,* November 19, 2020.

53 *DPCcars,* "Elon Musk Explains Tesla Motors Electric Vehicle History," June 12, 2016, video, 19:47.

Tesla's objectives to produce infinitely scalable clean energy generation and storage products.

The challenges the EV sector faces go beyond the battery technology research and development, and the capability to lower manufacturing costs through economies of scale. Without an optimal charging station infrastructure network, the establishment and growth of the EV market is simply not viable. As such, Musk explains how the launching of Tesla's first Model S in 2012 came in parallel with the deployment of the company's Superchargers network. Such infrastructure was critical to enable high-speed charging for solving the long-distance travel problem and, in general, for providing drivers the freedom to move everywhere.

"They [EV charging stations] were fundamental to really answer the question of 'can I drive my car long-distances?' And what it really comes down to is freedom. When you are buying a car you are really buying freedom to go where you want to go, and if you are constrained to a charging location you don't have the freedom. Superchargers are about giving you the freedom and making it really easy and convenient to go wherever you want."

Musk's point on the necessity for a robust EV infrastructure network reminded me of my interview with Georgetown McDonough School of Business adjunct professor Damian Saccocio. Saccocio shared with me his journey driving his Chevy Bolt EV in upstate New York. He expressed the inconvenience of having to wait in line for other people to charge their cars first. Likewise, he admitted he wasn't aware of the extended time it would take for his Bolt to charge.

"We rented a lake house and I couldn't get my Bolt up there, at least not reasonably. I hadn't really focused on how fast it recharged, but it turned out that was important if I want to take it on long distances. I couldn't count on pulling in and charging it."

"If somebody was ahead of me, or three people were ahead of me, the uncertainty of waiting two to six hours in the end made the solution not work for me for that use case. So I thought it was a good reality for a second car in the city. It was perfect for that other use case because we don't drive that far."

Saccocio's EV experience clearly sent the message there is still a gap to be closed in the US EV charging station infrastructure system. After all, the mission of Tesla and other EV makers is to serve people's various mobility needs, not urban transport only.

We have to recognize, nonetheless, the rapid expansion of public and private EV charging infrastructure networks nationwide. It went from roughly twelve thousand charging outlets in 2012—the release year of Tesla's first Model S—to nearly one hundred thousand in early 2021.[54,55] Michael Berube refers to this point in our interview, and considers it a solvable issue as more and faster EV charging stations get deployed across the US.

54 Jeff Plungis, "How the Electric Car Charging Network Is Expanding," Consumer Reports, last modified November 12, 2019.

55 I. Wagner, "Electric vehicle charging stations and outlets in US - February 2021," Statista, February 16, 2021.

"If you look today, fast charging infrastructure has been put along highways already. Between what Electrify America [US EV public charging network] has done, and other individually independent charging networks, and certainly what Tesla has done, you can solve that problem."

Likewise, technology companies like Silicon Valley-based StreetLight Data are helping with EV infrastructure planning by leveraging technology to map existing vehicle charging locations and inform where to deploy new capacity.[56]

Berube suggests the aforementioned battery technological advancements are enabling longer EV's travel ranges, and thus also helping to mitigate the apparent persistent EV charging infrastructure problem.

"Charging stations will not prevent electric vehicles from being successful."

He considers if we can get the range people need in the three-hundred-mile range, which is where we are right now, it changes the whole charging perspective.

So, what is the average EV travel range? You may be wondering. Every car in Tesla's portfolio, for example, has over three hundred miles of range, and the Tesla Model S offers a four-hundred-two-mile travel range, significantly more than any other EV ever produced. Other vehicles including the Chevrolet Bolt, Kia Niro EV, and Hyundai Kona Electric have a range of roughly two hundred fifty miles and overall,

---

56   "EV Infrastructure Planning," StreetLight Data, accessed August 9, 2020.

most EVs can travel well over two hundred miles before needing to charge.[57] Speaking to the rapid evolution of battery technology, Tesla's CEO Elon Musk said in November 2020 the automaker is working on achieving six hundred twenty-one miles of range for future vehicles.[58]

These fast improvements in EV travel range suggest the deployment of an effective charging station infrastructure network will very soon be more of a historical barrier and no longer as big of a consideration moving forward.

Another point Berube brings up is in regard to residential EV charging stations and the different power levels available. He illustrates the convenience of being able to charge the EV at home according to the users travel behavior, and considers that although currently expensive, Level 2 chargers can serve the purpose and will eventually come down in price. He suggests Level 2 charging can be done overnight, and we need to look at the trips people take and how long these trips take.

Reflecting on this, I think it would be a lot to charge three hundred miles overnight. Most individuals, however, wouldn't come home with a near-to-zero battery level. The vast majority of the time, drivers will probably go back to their homes with one hundred fifty miles, meaning they can plug in overnight and be topped off. More so, people without the need to drive long distance may return with two hundred

---

57   Steven Loveday, "Electric Cars With the Longest Range in 2021," US News & World Report, September 16, 2020.
58   Roberto Baldwin, "Tesla Is Working on 620-Mile Range for Future Cars, Upcoming Semi," *Car and Driver*, November 24, 2020.

fifty to two hundred seventy miles of range. Berube speaks to this topic:

"Most people will be filling in maybe seventy to eighty miles of range that they don't have to plug in every night, but you would have to have a Level 2 charger."

As a side note, Level 2 will charge a typical EV at a rate between twelve to sixty miles of range per hour, compared to only four to five miles for a level one charger.[59]

INCLUSIVE AND SMART CHARGING

According to Berube, the next and main challenge toward the full deployment of EVs in the US is making sure we have electrification available for people at all income levels; for minority and low-income communities might already be in areas that don't have as much infrastructure invested. Similarly, he refers to communities where people are in multi-unit dwellings, or in single home, duplex-type arrangements where they have no garage and they're on street parking, essentially.

"And how do you serve those people? It's great if you have your own place to park on your property. But for all the people who don't have that we have to solve that problem."

Berube, however, remains optimistic and suggests we have electricity everywhere, and it's about getting that charging

---

59 "Understanding the Different EV Charging Levels," *JuiceBlog (blog)*, Enel X, May 8, 2019.

point, the process, and about setting up the business models. As such, it will take some work and time over the next ten years when we will see far more charging everywhere. Even so, Berube claims it would take longer for all vehicles on the road to be electric.

To Berube's argument, could we have 90 to 100 percent of vehicles be electrified by 2030? I would say not without a mandate. However, that's for new vehicles only. In other words, that still means only 5–10 percent of all vehicles on the road would be electric by 2030. From that point on, I consider, it would take ten to fifteen years for all vehicles to be there.

In line with this, Berube believes we have time to develop the necessary EV infrastructure to accommodate a 100 percent electric fleet.

"We need to build the infrastructure, but we don't need to build it all in the next two or three years. We have ten plus years to build that out."

I certainly envision a significant increase in electric fleet over the next two decades. To achieve this more sustainable future, however, it will come to the continuation of rapid advancements in battery technology, charging infrastructure, and regulatory support.

Policy makers and the federal government are undoubtedly called to play a key role here. We've seen in the past positive outcomes from EV mandates, and associated federal subsidies and incentives toward the deployment of electric vehicles.

Another potential breakthrough in EV charging technology Berube mentions is the idea of smart charge management with the electric grid. He explains from an environmental perspective the promise of electric vehicles of low greenhouse gas emissions works if there is a reliable and resilient electric grid.

"You need to have renewables in a response-sensitive system, both from the demand side as well as the supply side."

Smart charging, in fact, refers to a charging system where EVs, charging devices, and charging operators (e.g., electric utility companies) share data connections.[60] Contrary to traditional charging devices, smart charging enables the charging station owner to monitor, manage, and restrict the use of their devices remotely to optimize energy consumption.

As such, utilities can control when they send electricity to EV owners based on their real time needs, avoiding peak demand levels. For example, consider a customer saying he or she needs to be charged at 8:00 a.m. the following morning. The utility could say they will not send the user any electricity until 2:00 a.m. when all the other loads are down. Likewise, they could send electricity to someone else at 3:00 a.m. because they need less. Utilities would know how much juice each EV needs.

In short, smart charging management would certainly enable the grid to become more flexible while responding to EVs electricity needs.

---

60 "Smart Charging of Electric Vehicles," Virta, accessed January 11, 2021.

MAKING A CASE FOR ELECTRIC COMMERCIAL VEHICLES

Toward the end of our interview, Michael Berube claims over the next five or six years we're going to see a huge push on commercial vehicles to be electrified. We started out with personal lighter vehicles because the thought was that batteries would be too big and heavy, and we wouldn't be able to get the range we need in larger heavy vehicles.

Yet, I think of companies like FedEx, UPS, Walmart, and PepsiCo with large captive fleets as they are all realizing EVs have actually a lower total ownership cost. The maintenance cost can, in fact, be lower than the operating cost per mile for fuel.

Berube thinks we're going to see in the next five years an explosion of electric commercial vehicles driven by economics, and some practical realities that will further help drive the volume of batteries and infrastructure.

"And I think you're going to see commercial vehicles surpassing the personally owned vehicles."

I do in fact believe deploying a network of more efficient electric buses with lower operating costs would result in cities having more of these in fleet and thus reducing wait times for buses. This would translate into a more accessible and reliable public transit system for the underserved communities. A more efficient transit system would also attract more high-income commuters, and thus tackle the traffic congestion problem by taking private vehicles off US roads.

Aside of the aforementioned economic benefit companies could see by electrifying their fleets, there is an environmental responsibility incentive that keeps gaining momentum in the US. Indeed, even though heavy-duty commercial vehicles comprise only about 5 percent of all vehicles on US roads, they are responsible for over 25 percent of greenhouse emissions coming from the transportation sector.[61] As such, a number of electric truck and van companies including BYD, Daimler, Nikola Motors, Tesla, and Rivian are putting increasing efforts toward the production and deployment of electric heavy-duty vehicles.[62]

Likewise, in 2019 e-commerce giant Amazon purchased one hundred thousand electric vans from startup Rivian Automotive. In fact, as part of the company's climate plan to be carbon neutral by 2040, Amazon expects to deploy ten thousand vehicles as early as 2022 and a full fleet of one hundred thousand by 2030.[63]

This push toward a reduction in carbon emissions through the electrification of commercial trucks reminded me of my consulting experience with the Hinduja Group from India. The Indian conglomerate's automotive subsidiary Ashok Leyland had a competitive positioning within the electric mobility space, and precisely in the manufacturing of electric buses.

61 "Cars, Trucks, Buses and Air Pollution," Union of Concerned Scientists, last modified July 19, 2018.

62 Shane Downing, "8 electric truck and van companies to watch in 2020," GreenBiz, January 13, 2020.

63 Annie Palmer, "Amazon Debuts Electric Delivery Vans Created with Rivian," CNBC, last modified October 8, 2020.

Recognizing India's concerning environmental issues, Ashok Leyland was determined to play a key role in reducing the exorbitant levels of air pollution in the nation. Similarly, recalling Berube's statement on battery being the key component of EVs, I recall we had our team counterpart centering their investigation on battery technology and strategic initiatives for battery manufacturing.

On this same topic, I came across a *Mobility* podcast episode where Ryan Popple, cofounder and executive director of Proterra, the leading innovator of zero-emission, battery-electric buses, explains how his military background led to his interest in reducing oil dependence.[64] As such, he developed a passion toward the adoption of EVs.

"My journey to electric vehicles started in 2006...and I was very much interested in working on something related to energy security and the future of energy. Quite a bit of that was driven by spending four years on active duty in the US Army, thinking what we're currently doing is not strategically sustainable."

"How much we take for granted the fact that the world effectively needs one hundred million barrels of a specific commodity every day, or things start stopping."

Popple continues explaining how he became interested in the alternative energy space, but with a focus on transportation. As such, he remembered attending a seminar at the

---

64 Ryan Popple, "#055: Electric Buses 101 with Ryan Popple, CEO, Proterra," October 18, 2019, in *Mobility Podcast*, produced by Greg Rogers, podcast, Soundcloud, 40:10.

University of California, Berkeley, and was fascinated about the fact there already was an existing industry and infrastructure that delivered electricity and could facilitate the transition to electric vehicles.

"One of the problems [with biofuels] is there really isn't biofuel distribution infrastructure...what I think is so fascinating about electric cars for the first time you can change that, and you also have a big enough industry, the electricity industry to somewhat counterbalance the influence of the oil and gas industry."

Although Popple expresses his motivation of electrifying every vehicle type and mode of transport, he claims economic priority and the price of electric vehicle technology needs to be considered. He says the bus segment offers a market with a viable business case and operating model. Popple, in fact, suggests the characteristics of how transit buses operate fit really well with EVs technology and EV charging today. He points out the use of diesel involved in running buses in the US every year are incredibly high.

"They're lucky to get four miles per gallon on diesel. They're running almost all the time in big cities, so you're looking at forty thousand miles a year, and that turns out to be ten thousand gallons of diesel per year, and per bus. So, every mile that a diesel bus goes down the road, it burns a quarter gallon diesel and emits it as evolution. So it's a hyper user of fuel."

Furthermore, Popple explains how EVs will be the long-term solution, but for this to happen people need to understand the economic value proposition. While a battery electric bus is currently 50 percent more expensive upfront than a diesel bus, five years ago it was 100 percent more expensive, indicating the rapid cost down curve. The 50 percent premium price, would be paid back with energy savings, spare parts, and maintenance.

"The most specific example we have so far is in brake systems. An electric vehicle regenerates energy with the electric motor. It doesn't touch the mechanical brakes unless it absolutely has to. So, what we're seeing is the durability of the braking systems in our buses, vastly exceeding internal combustion engine vehicles. It's a very expensive process to change out discs and pads on an internal machinery bus."

In addition, Popple discusses the importance of establishing partnerships between EVs and utility companies. The transportation, energy, and the electricity sector are coming together, and fleet electrification adds more value to utility suppliers than cars do.

Likewise, Popple shares the time when he testified in front of Congress to explain how we should approach the challenge of energy diversification, and to focus on electric vehicles as an economically viable market, allowing the industry to scale and improve its cost structure. As such, the transit bus market, school bus market, and delivery trucks are growing by orders of magnitude.

Similarly, the EV advocate criticized how in the US we are too focused on cars, to the point there exists a $25 billion loan guarantee program within the Department of Energy that can be applied toward light duty vehicles only. Popple, therefore, suggested to Congress to expand the current electric vehicle focus to include the commercial vehicle sector. As such, he requested Federal Transit Administration (FTA) support by restructuring their procedure to allocate funding for transit projects.

"You have to use about 50 percent of the capital expenditure in general. We propose that funding be calibrated based on the energy efficiency of the technology deployed because if you are subsidizing a city or county, you would naturally want to provide them with technologies and products. They're going to have lower ongoing operating costs, so they don't have to come back for more funding. It's the responsible thing to do."

In short, Popple argues the idea that electric vehicles and, more specifically electric buses, represent a crucial market and sector US agencies and politicians need to put efforts on. Indeed, the goal is to gradually stop depending on traditional combustion vehicles, and thus alleviate the pressing air pollution problem.

THE FUTURE IS ELECTRIC
Although the electric share of total vehicle sales in the US is still small at about 2 percent in 2019, different factors are driving the electric markets growth.[65] Advancements in battery technology and charging infrastructure, a gradual

---

65 "How Many Electric Cars Are On The Road In The United States?," USA Facts, last modified October 22, 2020.

decrease in battery cost production, governmental policy incentives to lower emissions, and increasing investments in electrification by automakers and large fleet operators are set to accelerate the EV market maturity.[66]

Moreover, I foresee other potential innovative developments around EVs. Consider dynamic wireless EV charging eventually happening beneath city roads while self-driving shuttles are driving down the street.

I am confident the electric vehicle industry will keep transforming the US mobility sector over the next few decades. Policy makers are called to keep promoting the implementation of more environmentally friendly mobility solutions.

The private and public sectors should ensure inclusive electrification for those who wish to be served. Furthermore, electric buses are well positioned to facilitate more accessible and reliable public transportation to the underserved population and attract high-income commuters, thus reducing traffic congestion levels.

KEY TAKEAWAYS
- As transportation remains the largest source of carbon emissions in the United States, electric vehicles (EVs) have appeared as a low-carbon alternative to gasoline-powered vehicles.

---

66 Colin McKerracher et al., "Electric Vehicle Outlook 2020," BloombergNEF, *Bloomberg*, 2020.

- Federal policy and government incentives toward EV solutions, the rapid developments in battery technology, EV charging infrastructure network, and smart charging management are enabling the growth of the EV market.
- The lower ownership cost for EVs compared to traditional combustion engine fleet are pushing original equipment manufacturers (OEMs) and fleet operators to produce electric commercial trucks, further enhancing the EV industry.

CHAPTER 4:

EMERGING TECHNOLOGIES FOR A BRIGHTER FUTURE

"There is no alternative to digital transformation. Visionary companies will carve out new strategic options for themselves—those that don't adapt, will fail. The number of digital options open to businesses is growing exponentially and with no sign of letting up."[67]

—*JEFF BEZOS, AMAZON CEO*

Emerging technologies will play a crucial role in our ability to improve the future of transportation.

Some of these trends that should excite investors, policy makers, commuters (and that certainly excite me) are the internet of things (IoT), 5G, cloud/edge computing, big data,

---

67 Jeff Bezos, quoted in Express Information Systems, "Four Key Trends Reshaping Wealth & Asset Management Accounting," November 9, 2020.

and artificial intelligence (AI). During my time at Wipro Limited, a leading global technology, consulting and business process services company, I've closely worked at the intersection of some of these cutting-edge technologies and the manufacturing and automotive/mobility sectors. As such, I've witnessed how the digital wave is transforming multiple industries.

These technology trends are indeed revolutionizing the way everything works in the world and ruling the transportation industry. The integration of these technologies in new mobile app development has been a game-changer for the capabilities of services in the transportation space, and for the sheer number of people these services can reach.

These ubiquitous technologies are a part of our daily lives, running everything from our smartphones to our microwaves and ovens. In the following chapters, we'll delve into how they are transforming the different areas of the transportation sector, but before we do let's take a moment to describe each of them at a high-level.

INTERNET OF THINGS (IOT), 5G, AND CLOUD/EDGE COMPUTING

In the broad sense, the term internet of things (IoT) is an ecosystem of connected physical devices to the internet, and it is increasingly being used to define objects that "talk" to each other. An article from *WIRED* magazine, authored by

Matt Burgess, collects various definitions and perspectives on IoT from various technology experts.[68]

According to Matthew Evans, IoT program head at techUk, UK's leading technology membership organization:

"Simply, the internet of things is made up of devices—from simple sensors to smartphones and wearables—connected together."

Caroline Gorski, the head of IoT at Digital Catapult, a UK agency for the early adoption of advanced digital technologies, suggests IoT allows devices to communicate with others on closed private internet connections:

"It's [IoT] about networks, it's about devices, and it's about data...the internet of things brings those networks together. It gives the opportunity for devices to communicate not only within close silos but across different networking types and creates a much more connected world."

In short, IoT allows the collection and analysis of data and information to generate insights and take action toward what we want to achieve or solve. IoT democratizes decision making for all devices. It's a move away from a world that requires human interaction to configure devices to behave a certain way to one where devices talk to each other and can decide on their own (or can receive remote instructions) for how to behave.

---

68 Matt Burgess, "What is the Internet of Things? WIRED Explains," *WIRED*, February 16, 2018.

To illustrate the power and benefits of IoT, let's consider the healthcare sector. For instance, data collected from continuous positive airway pressure (CPAP) machines used to treat sleep apnea—a potentially serious sleep disorder in which breathing repeatedly stops and starts—can draw a clearer image of what actions a person needs to take to decrease the severity of their sleep apnea. With the inclusion of IoT in such devices, doctors can better monitor and analyze the sleep patterns of their patients remotely, such as horizontal and vertical positioning, and it can help them identify ways to lessen the effects of the sleep apnea in their patients' lives.[69]

I was personally diagnosed with sleep apnea many years ago, and since then my CPAP machine and I have become inseparable friends. Yet, my "old-fashioned" device is, unfortunately, not IoT-enabled and comes with an inserted SIM card instead, which I have to take with me to my doctor appointments so the stored data can be analyzed. Don't ask me why but I have only scheduled one follow-up appointment since my initial sleep apnea diagnosis, and I cannot tell whether the breathing in my sleep has improved or worsened. Considering my negligence at routinely visiting my doctor, perhaps I ought to upgrade my CPAP and have him contact me should he remotely identify ways to improve my sleep.

The above is only one example of the gradual importance and advantages IoT is bringing to our lives. As you may now rightfully imagine, through the inclusion of distinct tools

---

69 "How IoT Is Changing Sleep Therapy," *Blog*, Aeris, accessed September 8, 2020.

and services, the adoption of IoT has also led to tangible benefits to the transportation sector, in particular through vehicle telematics.

Telematics integrate the utilization of telecommunications and informatics for vehicle applications. It's the tracking, monitoring and connectivity of vehicles. Examples of telematics use cases include fleet management, where an IoT-enabled diagnostics solution can help drivers and fleet managers be in control of vehicle maintenance issues before they can lead to a potential breakdown. For instance, the vehicle will regularly send engine diagnostics back to head office so maintenance can be scheduled ahead of any engine problems.

Moreover, through traffic congestion control systems, reservation and booking systems used by operators, toll and ticketing, and security and surveillance, IoT is increasingly transforming the transportation space.[70] Similarly, IoT allows the implementation of remote vehicle monitoring systems, providing an improved traveler experience, increased safety, and enhanced operational performance. For example, an IoT-powered vehicle tracking system allows authorities to check the driver's route and driving habits, monitor how fast the vehicle is moving, and take necessary actions.

According to a study report by Synarion IT Solutions, global IoT in the transportation market valued at $135.35 billion in 2016 is expected to reach $328.76 billion by 2023 and is

---

70   "IoT in Transportation Market—Growth, Trends, Covid-19 Impact, and Forecasts (2021 - 2026)," Mordor Intelligence, accessed September 9, 2020.

growing at a compound annual growth rate (CAGR) of 13.1 percent from 2017 to 2023.[71]

We cannot talk about IoT without mentioning 5G (5th generation mobile network). After all, we referred earlier to IoT as an ecosystem of connected physical devices to the internet. As such, 5G connectivity is expected to be one of the fastest wireless technologies ever created.[72] Consider, for example, how you use smartphones on 4G LTE (Long Term Evolution) networks today: messaging or video-calling your friends and family, playing games, and streaming and downloading You-Tube videos and Netflix shows. The 5G wireless technology is capable of delivering much faster data upload and downloads, up to ten times faster in early tests.[73]

Moreover, 5G technology is expected to greatly enhance the transportation industry. Telecommunications company AT&T highlights:[74]

"When added to existing network architectures, 5G technology has the potential to provide increased visibility and control over transportation systems."

This 5G brings the potential to provide end-to-end connectivity across cities and beyond. Transportation companies will be able to leverage 5G for their fleet to communicate

---

71 "Impact of AI, IoT And Big Data On Transportation Industry," *Blog*, Synarion Solutions, accessed September 9, 2020.

72 "What Does 5G mean?," Personal Tech, Verizon, November 5, 2019.

73 Caitlin McGarry, "5G vs 4G Performance Compared," Tom's Guide, February 25, 2021.

74 "How 5G Will Impact the Transportation Industry," AT&T Business, AT&T, accessed September 9, 2020.

in a multifaceted ecosystem: vehicle-to-vehicle (V2V), vehicle-to-infrastructure (V2I) (e.g., sensors on bridges, roads and traffic lights), vehicle-to-pedestrian (e.g., people's smartphones), and vehicle-to-network (V2N). Likewise, as self-driving vehicles become a reality, these vehicle communications will lead to safer and more reliable autonomous driving algorithms.

The 5G market is projected to increase at a whopping compound annual growth rate (CAGR) of 43.9 percent from 2021 to 2027.[75]

Cloud and edge computing are other emerging technologies enhancing IoT devices. Nima Negahban, chief technology officer at Kinetica, explains the difference between the two:[76]

"Edge computing is about processing data locally, and cloud computing is about processing data in a data center or public cloud."

Dr. James Stanger, chief technology evangelist at CompTIA, illustrates how networks are transitioning from cloud to edge computing for connecting IoT devices:[77]

"As IoT connects more and more devices, networks are transitioning from being primarily highways to and from a central location to something akin to spiderwebs of interconnected,

---

75 "Global 5G Services Market Size Report, 2021-2027," Grand View Research, May, 2020.

76 Stephanie Overby, "How to Explain Edge Computing in Plain English," The Enterprisers Project, November 30, 2020.

77 Ibid.

intermediate storage, and processing devices… The data is stored at intermediate points at the edge of the network, rather than always at the central server or data center."

By keeping data closer to end users, latency becomes less of a problem with edge computing. It brings the advantage of processing time-sensitive data while cloud computing is used to process data that is not time-driven. As such, businesses would prefer edge computing in cases where a delay in the machines decision-making process due to latency would result in losses for the organization.

Edge computing is also enhancing the transportation industry with more connected and intelligent solutions. For instance, transit operators are able to decrease passenger waiting time by taking real-time decisions. Public bus operators are able to improve service levels and passenger experience by tracking their fleet and through live video monitoring.[78]

According to the world's leading research and advisory company Gartner, while 90 percent of all data is created and processed inside traditional centralized data centers or clouds in 2020, 75 percent of data will be processed through edge computing by 2025.[79]

BIG DATA

Gartner defines big data as:

78 "Transportation Takes A Leading Edge with Smart Technology," *TechHQ*, July 21, 2020.
79 Ibid.

"Data that contains greater variety arriving in increasing volumes with ever-higher velocity."[80]

Big data integrates data from several sources and applications, is stored and managed, and analyzed. It's typically stored in dedicated data centers. The data set is typically too big for a single machine and spans many (sometimes hundreds) of physical machines. The difficulty lies in how to efficiently process that data by deciding what you don't need for a specific question/query early on and eliminating that from your search path. Advanced statistical analysis concepts play a major role.

Interestingly enough, the data scientist profession was born from the big data movement and there are now degree programs for it.

In simple words, big data is larger, more complex data sets, particularly from new data sources, which are not able to be managed by traditional data processing software due to its immense size. These voluminous data can, nonetheless, be utilized to address business problems impossible to tackle before.

For instance, companies like Netflix and Procter & Gamble use big data for product development. They anticipate customer demand. They build predictive models for new products and services and analyze the relationship between the commercial success of these and key attributes of their past and current offerings. Moreover, big data is leveraged

---

80   "Big Data Defined," Oracle, accessed September 9, 2020.

to improve customer experiences. It enables companies to collect data from distinct sources including websites, social media platforms, and call logs to improve client interaction and maximize the value delivered.

Over the past few years, big data has transformed a myriad of industries and transportation is not the exception.

To exemplify the large volumes of data generated, consider INRIX's Global Traffic Scorecard mentioned in the second chapter of this book. INRIX calculated their results by combining anonymous, real-time global positioning system probe data from three hundred million connected cars and devices with real-time traffic flow data and other criteria, including construction and road closures.[81] Yes, three hundred million vehicles!

Synarion's report predicts the big data market will increase from $42 billion in 2018 to $103 billion in 2027.[82]

The emergence of big data combined with the explosion of data coming from mobile devices has disrupted the transportation space. Leveraging technology, mobility has gained strong momentum which leads me to be hopeful and optimistic about the future of the fragmented US transportation system.

---

81 Trevor Reed, "INRIX Global Traffic Scorecard," INRIX, INRIX Research, March, 2020.

82 "Impact of AI, IoT And Big Data On Transportation Industry," *Blog*, Synarion Solutions, accessed September 9, 2020.

ARTIFICIAL INTELLIGENCE (AI)

Investopedia, the world's leading source of investing and finance education, refers to artificial intelligence (AI) as:[83]

"The simulation of human intelligence in machines that are programmed to think like humans and mimic their actions. The ideal characteristic of artificial intelligence is its ability to rationalize and take actions that have the best chance of achieving a specific goal..."

"...is based on the principle that human intelligence can be defined in a way that a machine can easily mimic it and execute tasks, from the simplest to those that are even more complex. The goals of artificial intelligence include learning, reasoning, and perception."

Chances are most of us have experienced AI one way or another in our daily lives, and much more frequently than what we can imagine.

Consider face detection and recognition, using virtual filters on our face or using face ID for unlocking our phones. How does this work? Smart machines are taught to identify facial coordinates, landmarks (e.g., nose, mouth, eyes), and alignment. In more detail, machines are taught how to learn so they can adapt to meet their goal. A programmer would originally teach machines how to decide what a facial feature looks like, and how to adjust if they would get it wrong. Then the human would train the AI with a large dataset. The AI

---

83 *Investopedia*, s.v. "Artificial Intelligence (AI)," by Jake Frankenfield, accessed September 9, 2020.

system would start failing early on, but then without human interaction would be able to tell it was failing and make the necessary adjustments. As time goes by, its success rate will increase to nearly perfect if all goes according to plan.

Similarly, think about the times you have made use of a company's website chat-bot before calling or emailing them, receiving a quicker and more effective solution to your inquiry. Further, chances are you have noticed how the Microsoft Word document you are typing identifies incorrect usage of language and suggests corrections. This is possible through an artificial intelligence algorithm using machine learning.

Other examples of AI applications include search and recommendation algorithms when seeking for favorite movies on Netflix, or when calling on your Amazon Alexa speaker to play music. Whether you love it and find fascinating how well these apps know your preferences, or whether you see it as a scary intrusion to your mind, truth is AI's capabilities keep rapidly evolving.

The rise and power of AI is recognized by US business and political leaders:[84]

"Software is going to solve, where it'll look at all the new information and present to you knowing about your interests, what would be most valuable. So, making us more efficient."

—BILL GATES, FOUNDER, MICROSOFT

---

84 *Simplilearn,* "Artificial Intelligence & the Future—Rise of AI (Elon Musk, Bill Gates, Sundar Pichai)|Simplilearn," March 26, 2019, video, 4:51.

"It is [AI] seeping into our lives in all sorts of ways we just don't notice. We're just getting better and better at it and were seeing that happen in every aspect of our lives, from medicine to transportation to how electricity is distributed, and it promises to create a vastly more productive and efficient economy. And, if properly harnessed, can generate enormous prosperity for people, opportunity for people, can cure diseases we haven't seen before, and make us safer because it eliminates inherent human error in a lot of work."

—FORMER UNITED STATES PRESIDENT BARACK OBAMA

"AI is probably the most important thing humanity has ever worked on. I think of it as something more profound than electricity or fire."

—SUNDAR PICHAI, GOOGLE CEO

Similarly, AI has penetrated the transportation industry through a myriad of applications including self-driving/autonomous vehicles (AVs), traffic management solutions, smartphone apps, and passenger transportation. More specifically, consider how often you use GPS navigation software apps such as Google Maps or Waze when driving from one place to another. Sensors and cameras embedded on roads collect a voluminous amount of traffic data which is sent to the cloud, where it is analyzed by an AI-powered system.[85]

From there, valuable traffic prediction insights are gathered and provided to us commuters. We are notified of the shortest and best route to our destination, avoiding road blockages,

---

85 Naveen Joshi, "How AI Can Transform The Transportation Industry," *Forbes*, July 26, 2019.

accidents, and traffic congestion. As such, AI is being used not only to reduce undesired traffic but also to enhance road safety and shorten wait times.

I have to admit since the deployment of smartphones and associated traffic/GPS apps, I have become really bad with directions. My reliance and dependence on such satellite-based navigation tools has become quite extreme. Shame on me, not on AI.

According to Synarion IT Solutions report, the global AI in transportation market size was $1.21 billion in 2017 and is expected to reach $10.3 billion by 2030, growing at a CAGR of 17.87 percent.[86]

CASE STUDY: STREETLIGHT DATA IN TRANSPORTATION PLANNING

In a BrightTALK Founders Spotlight interview, cofounder and CEO of StreetLight Data, a data and software company mixed together, Laura Schewel explains how big data can change the way we plan and shares the origin and challenges that arose in creating the company.[87]

"We're the first and only company in our industry to mix together the explosion of data coming off of mobile devices with software to make that data useful for the transportation industry."

---

86 "Impact of AI, IoT And Big Data On Transportation Industry," *Blog,* Synarion Solutions, accessed September 9, 2020.

87 *BrightTALK,* "Founders Spotlight—Episode 6: Laura Schewel, Founder and CEO, Streetlight Data," May 9, 2019, video: 44:35.

Streetlight Data works with different app companies and cloud and IoT connected car and truck companies to collect "anonymized" data from mobile devices. The software determines how the mobile devices and the people connected to them inside the vehicle are moving around. This aggregate data is analyzed with the purpose of providing effective and efficient transportation planning.

Schewel has a strong motivation toward helping alleviate climate change. Her early professional career as a transportation sustainability person confirms this. In 2006, she started working at Rocky Mountain Institute, a think tank in Colorado, on electric vehicle deployments and planning for electric vehicles. This coincides with the time of the new big wave of electric vehicles with Tesla just coming out, and an increasing interest by a myriad of companies to gather data aimed to model the electrical grids planning. Nevertheless, Schewel suggests there was not much data on this.

"And what became really clear was everybody was asking this question that said, okay, get some data and then let's model how these electric vehicles might hit the electrical grid, or how they might show up and want to charge in grocery store parking lots. And there wasn't much data there."

Schewel's career journey continued at the Federal Energy Regulatory Commission in Washington, DC to work on the same idea of electric vehicle adoption from the perspective of the grid. Her clients wanted to have detailed models that would provide information on where electric vehicles needed to charge and how much electricity was needed. But again, as Schewel confesses, the data wasn't there. Although initially feeling discouraged, she saw this as an opportunity in the electric vehicle infrastructure space. It led her, in fact, to go back to graduate school and develop an app that collects data by leveraging the existence of GPS in all phones people carry around.

"I kept on looking and looking at it [data] and it just wasn't there. I identified that as a gap for electric vehicle infrastructure planning. I then went back to grad school at UC Berkeley and the Energy and Resources Group. I thought, hey, maybe this is research worthy."

"And that was around the time when Android and iPhone devices had started the backgrounding feature, which means they could run two apps simultaneously, not just the app that was in the foreground. And I thought, 'hey all these phones people are carrying around, they have GPS.'"

"Maybe I can develop an app that collects data. And that could be a feed for some of these electric vehicle, big picture questions."

Amidst the major financial crisis in 2008–2009 and the fear of seeing her PhD project funding compromised, Schewel decided to seek other ways to fund her lab and entered a business plan competition at the UC Berkeley Haas School of Business. Not only did she win a prize of $40,000–$50,000, but the competition judges, who were also investors, encouraged Schewel to bring her powerful idea to business reality. Even though she did not have initial intentions of starting a business, she decided to pursue this opportunity to make a larger environmental impact outside academia.

"And I first laughed them off, but I'd been thinking a lot about climate change and the frustration of working on climate change in an academic environment where all you get to do is write papers, and then worry that no one reads them."

"So I thought, I don't know, I'll take some of these [investors] meetings, and I started talking to them and looking more into the way I could have an environmental impact on transportation."

After over a year of focusing her startup business on collecting data for electric vehicle (EV) charging infrastructure planning, Schewel confesses companies in this space (e.g., car makers, EV charging stations) were not willing to pay for her business offering. This led her to think about the first conversation she had about gas station planning and repair facility planning, a form of transportation infrastructure. As such, she realized the data her company was gathering was critical for planning much more than solely EV infrastructure.

"Then we realized there is actually an enormous market and there is a ton of people who want to pay us. And that's when we could see a business model that actually had legs."

Furthermore, Schewel explains how after conversations with venture capitalists (VCs), she understood trying to develop her own app to collect data and monetize it was not an optimum business model. She instead partnered with connected vehicle companies that collect data for their own purposes, and which immediately gave her access to millions of mobile devices.

"And right now we're analyzing over seventy million devices every day. We could have never gotten that kind of data trying to make our own app."

Schewel describes the two above-mentioned moments as key for StreetLight Data to shift its business model. From an overly niche electric vehicle market with its own app, the startup pivoted to an incredibly powerful business model of taking advantage of data coming off of thousands of other people's projects and businesses. Now the company harnesses such data for a generalized transportation infrastructure software.

"So we developed a product that answers a direct and clear need for the transportation community by listening to the community and understanding exactly what they needed, so they will buy."

Today, StreetLight Data sells a software platform that allows the company's clients across transportation agencies, engineering firms, and private transportation companies to log in and look up a transportation behavior at a specific place—a specific intersection of a specific town anywhere in the US and Canada.

For example, when planning a new bus route across the Golden Gate Bridge in San Francisco, the agency may seek to know where the bus stop should be to get most people conveniently into a specific area. Thus, with StreetLight Data you can light up the Bay Bridge and identify the neighborhoods where most people originate and where most people are going on a typical month. By using that map, the

StreetLight Data's client can gain more focus and efficient transportation planning.

Schewel describes StreetLight Data offering as a product for transportation planners. The company looks on decisions that have a lead time—next five or ten years—and these decisions are usually around infrastructure and therefore large investments. For instance, a $4 billion highway pension, a medium project like a new bus route, or something small like the location of a stoplight at a particular intersection.

"We're always working on the system, the infrastructure, the things around you, and then we hope that makes it for the individual a better transportation experience. So, we're not influencing their decisions in a direct way, we're influencing the system in which they can make decisions."

Schewel highlights during a Tedx Talk the environmental impact of miles traveled by personally owned vehicles. She illustrates this through the case study of grocery shopping transportation behavior in Sacramento, California.[88]

"They [cars and trucks] contribute 30 percent of America's greenhouse gas emissions. And if you're a Californian as I am, it's probably 70 percent of your greenhouse gas emissions, and a third of those miles traveled are going to buy stuff like groceries, and that is ridiculous and is also scary."

---

88 *TEDx Talks,* "Drive for Change | Laura Schewel | TEDxSacramentoSalon," August 11, 2015, video, 11:00.

Schewel considers the environmental community is not paying too much attention to the environmental effects of miles driven. There is an enormous data gap when it comes to transportation. She shares a use case of how StreetLight data is helping bridge this gap. The company created a map illustrating all the trips that went to a Whole Foods in Sacramento, while identifying which trips were in personal cars versus commercial trucks that deliver the food. The software was able to determine what percentage of shoppers at the Whole Foods live in a particular neighborhood, from one mile to eight miles from the store.

Schewel suggests not only these AI-powered insights can be used to take transit decisions, such as the implementation of a bike lane or a bus driving through a specific route, but they can be leveraged to take business decisions. For instance, it could help decide the placement of a grocery store within a reasonable distance from those people living far from the Whole Foods they usually shop at. Besides serving a targeted market, this initiative would help alleviate the carbon emission issue by reducing personal car miles traveled.

"I'm more concerned with these people up here, 14 percent of the shoppers live eight miles away, and that's about a twenty-minute drive, which is pretty long for going to the grocery store. So, we might think, 'what are those people doing, maybe they really love Whole Foods,' or maybe there is not that many options for them."

"So here I see an opportunity to make money, and to save some carbon. Someone should put an organic fancy grocery store up there."

Overall, Schewel has found a way to fulfill her original mission of combatting climate change, from helping reduce vehicular carbon emissions to enhancing EV charging infrastructure. Better still, leveraging the power of big data and internal machine learning and AI capabilities, StreetLight Data has contributed to a multitude of other use cases across transportation planning. The company's obtained metrics are helping inform and manage transport's most complex challenges; helping distinct transportation entities bridge the existing gaps in this space.

STARTUPS TO SHAPE THE FUTURE OF MOBILITY

The success of StreetLight Data coming from the startup world and humble academic-focused beginnings reminds me of my interview with Georgetown McDonough School of Business adjunct professor Damian Saccocio. Saccocio teaches a technology strategy course which I had the opportunity to enroll in during my MBA. The class explained the role of technology in business strategy across numerous industries and illustrated patterns and cycles of technological development through a combination of theory, history, and case studies.

Saccocio is also VP of Management Leadership of Tomorrow (MLT) and focuses on steering investments in startups and innovation, technology, research, and analytics.

In our interview, Saccocio commented on the important and vital role startups play to explore different technological solutions, so we don't get stuck in dead ends or cul-de-sac technology.

"Startups have a huge role to play in hedging our technological bets."

In fact, Saccocio believes startups are going to find solutions to problems the incumbents think are not solvable or optimizable. He illustrates this thought through the example of fracking—the hydraulic fracturing technique into the earth to extract natural gas or oil.

"Even fracking, if you know the history of horizontal drilling, the incumbents all assumed it couldn't be solved, and it was a startup who really poked at these problems along the value chain and found breakthroughs."

"Suddenly horizontal drilling is possible and now the US is a net exporter of oil and Shell and Exxon did not do this. They came in afterwards."

According to Saccocio, the transportation sector is not the exception and startups will enhance the development of the industry.

"In transportation, some of these really hard problems we think are basically the physics or the science or the data, but I will look to startups to find the solutions."

KEY TAKEAWAYS
To put it in a nutshell:

- Emerging technologies such as the internet of things (IoT), 5G, cloud/edge computing, big data, and artificial intelligence (AI) are exponentially penetrating and transforming the transportation industry.

- Data coming from people's mobile devices is being leveraged for transportation planning, among other use cases.
- Innovative companies and startups are and will continue playing a key role in disrupting the transportation sector.

In the subsequent chapters of this book, I will take you through the journey of exploring distinct mobility trends empowered by the aforementioned emerging technologies. These mobility innovations include autonomous vehicles, micromobility, microtransit, shared mobility, and Mobility as a Service.

I am a firm believer these trends—in addition to electric vehicles covered earlier—can and should be leveraged together over the next few decades to empower the fragmented US transport system and drive mobility for all the population.

CHAPTER 5:

AUTONOMOUS VEHICLES

"Self-driving cars are the natural extension of active safety and obviously something we should do."[89]

—ELON MUSK, TESLA CEO

Traffic safety is and should certainly be a concern in the United States. According to the National Highway Traffic Safety Administration (NHTSA), there were 36,096 people killed and 2.74 million injured in motor vehicle traffic crashes on US roadways during 2019.[90] As such, road crashes are the leading cause of death in the US for people aged one to fifty-four.[91] In addition to the lives affected, these vehicular accidents carry a high economic cost. The Centers for Disease Control and Prevention (CDC) estimated the cost of medical care and associated productivity losses—from occupant

---

89 Will Oremus, "Tesla May Build Its Own Self-Driving Cars, But Prefers the Term Autopilot," *Slate*, May 7, 2013.

90 *Overview of Motor Vehicle Crashes in 2019*, National Center for Statistics and Analysis, Traffic Safety Facts Research Note, Report No. DOT HS 813 060 (Washington DC: National Highway Traffic Safety Administration, 2020).

91 "Road Safety Facts," Association for Safe International Road Travel (ASIRT), accessed January 4, 2021.

injuries and deaths—to be over $75 billion for motor vehicle crashes that occurred in 2017.[92] Unsurprisingly 94 percent of these serious crashes are due to human error.[93]

As discussed in the third chapter, emerging technologies such as artificial intelligence (AI), big data, and the internet of things (IoT), have been transforming the transportation industry. Ongoing advanced driver assistance systems (ADAS) already contribute to a reduction in vehicular traffic accidents. By helping drivers avoid making unsafe lane changes, warning them of other vehicles behind them when they are backing up, or even by having the automobile brake automatically if a vehicle ahead stops, artificial intelligence is already improving the safety of our roads.

The hope is with the continuous advancement of technology, fully automated vehicles (AVs) will become a reality one day and will drive us where we please, whether because we don't want to or we physically can't do it ourselves.

The Society of Automotive Engineers (SAE) defines six levels of automation, from no automation, where a fully engaged driver is required at all times, to full autonomy, where an automated vehicle operates independently without a human driver.[94] In more detail:

92 "Cost Data and Prevention Policies," Centers for Disease Control and Prevention (CDC), National Center for Injury Prevention and Control, last modified November 2, 2020.

93 "NHTSA: Nearly All Car Crashes Are Due To Human Error," *Ayers, Whitlow & Dressler* (blog), accessed September 10, 2020.

94 "Automated Vehicles for Safety," National Highway Traffic Safety Administration, accessed October 10, 2020.

"Level 0: No Automation—Zero autonomy; the driver performs all driving tasks."

"Level 1: Driver Assistance—Vehicle is controlled by the driver, but some driving assist features may be included in the vehicle design."

"Level 2: Partial Automation—Vehicle has combined automated functions, like acceleration and steering, but the driver must remain engaged with the driving task and monitor the environment at all times."

"Level 3: Conditional Automation—Driver is a necessity but is not required to monitor the environment. The driver must be ready to take control of the vehicle at all times with notice."

"Level 4: High Automation—The vehicle is capable of performing all driving functions under certain conditions. The driver may have the option to control the vehicle."

"Level 5: Full Automation—The vehicle is capable of performing all driving functions under all conditions. The driver may have the option to control the vehicle."

THE AUTONOMOUS EVOLUTION

While the concept of autonomous vehicles has gained momentum over the past few years, the truth is, conversations around self-driving cars have been around for a while.

The first truly autonomous cars appeared at Carnegie Mellon University circa 1980, according to open innovation platform for AI and Robotics Kambria.[95] Yet, it was DARPA's (Defense Advanced Research Projects Agency) self-driving car competitions in 2004, 2005, and 2007 that truly accelerated the evolution of AVs. DARPA is indeed the most prominent research organization of the US Department of Defense and has been promoting technology innovation within the AV industry and helped cultivate the new leading companies in autonomous technology.

DARPA challenges have been held across desert, urban, and mountainous environments. Automakers such as General Motors, Volkswagen, and Ford have partnered with engineering research institutions like Stanford University, Carnegie Mellon, and Virginia Tech for the development of their AVs for the distinct DARPA competitions.

In a rare opportunity to get firsthand insights into the AV industry, I conducted an interview with an expert from a leading worldwide automaker. Due to the highly competitive nature of the market, she requested to remain anonymous. One thing that stuck out for her was just how much the autonomous vehicle industry has evolved in recent years. Computations that used to require a trunk full of computers can now be accomplished with something as small as an iPad.

Form follows function, she says. The mere fact terabytes of data were a necessity resulted in awkward attempts to accommodate that computing power. Roof-mounted LiDAR [Light

---

95 "The History and Evolution of Self-Driving Cars," Kambria, June 23, 2019.

Detection and Ranging] cameras and other sensory equipment added to the experimental look of the vehicles back in 2005. Within ten years, those sensors and the computing power that powered them were seamlessly integrated into the iconic aesthetics of the vehicle. The power of these vehicles was amplified, yet the beauty remained, she told me.

We also talked about AV prototype experiments and other explorations in autonomous operation. They have sparked a range of reactions from consumers and employees. Sheer anxiety brought on by the impression of a loss of control was coupled with excitement at the prospect of hands-free operation associated with sci-fi movies. She described to me how some people would be totally *cool* and relaxed during their AV ride, while others would be literally holding on to the doors.

I began wondering how I would feel in a similar situation. The veracity of the popular phrase "you are in control of your own destiny" could certainly be challenged. In fact, I just didn't know how I would feel about having a "robot" taking full control of my vehicle during my journey. Would I also be grasping my door and praying?

Truth is I have always admired the remarkable advancements in robotics and automation technologies. I think of how I relate to automated machines, even to much more *basic* concepts like programmable coffee makers. Simply thinking about the idea to set the time I want my coffee to start brewing in the morning already puts a smile on my face. Programmable thermostats are another example of the perks of automation. Being able to run our furnaces more efficiently while ensuring we arrive to our homes at a comfortable temperature is a great benefit to have. Likewise, think of how going up and down an elevator has been for a very long time quite the norm, where we simply deposit, very often unconsciously, our faith in the automated rides they offer us.

In short, automation assists our lives in more ways than we know. It's aimed to make our lives easier from all sorts of different angles.

Interestingly enough, only a few weeks after my interview above I came across a Fairfax County (Virginia) press release announcing the deployment of an electric autonomous shuttle pilot project.[96] The self-driving vehicle Relay would offer people free rides on a fixed schedule between the Mosaic District—a residential, entertainment and shopping development—and the Dunn Loring Metro-rail.

The road-to-vehicle full automation is not simple and by now you might be asking yourself, what is the current level of self-driving vehicle technology? Aside from small scale AV pilot projects like Relay or Google's Waymo at Level 4, when it comes to mass AV production we have not passed Level 2. Examples of Level 2 AVs include Tesla's Autopilot and Nissan's ProPilot. Both vehicles can automatically keep the car in the right lane on the road and keep it at a safe distance from the car in front when in a traffic jam.[97]

---

96 "Fairfax County and Dominion Energy Launch Public Service on Virginias First Publicly Funded Autonomous Electric Shuttle Pilot Project," Fairfax County, October 22, 2020.
97 Chris Hall, "Self-Driving Cars: Autonomous Driving Levels Explained," Pocket-lint, August 19, 2020.

A REGULATED ENVIRONMENT

Autonomous vehicles face a complex matrix of federal and state regulation. Nevada was the first state to allow autonomous driving in 2011. As of 2021, a decade later, twenty-nine states plus Washington, DC have passed legislation on self-driving vehicles.[98]

My anonymous interviewee says there was a time when most states that allowed AVs required a car plate unique to their state. This implied when cruising from California to Nevada, for example, you would literally have to stop at the border, unscrew California's license plate, and put up the new license plate for Nevada that had the approval to drive autonomously.

To ensure AVs safety, vehicles need to communicate with each other and the infrastructure around them. LiDAR and cameras are being used to identify the markings on a highway and for the self-driving vehicle to understand in what lane it's driving. The challenge that persists is how to get off a highway. She believes if we want to become a truly AV-driven society, we need to start looking at the traffic infrastructure.

To accommodate the ways smart vehicles might communicate with both other machines and humans, new signals and protocols need to be developed. She shared a very interesting project she had been working on where her team tried to replicate what would typically happen when a car gets to a four-way stop. This refers to the experience we have all

---

98 Melanie Musson, "Which states allow self-driving cars?," Auto Insurance, last modified February 26, 2021.

had when arriving at an intersection simultaneously with another driver. You gesture with a wave of the hand, signal with a light or tap of the horn to communicate and say it's your turn, go ahead, or signal to a pedestrian it is okay to walk. This speaks to the challenge of having the vehicle gesture if there is nobody in there or you're just sitting in the backseat.

Another challenge for AVs could be in the case of a power outage and thus disabled traffic signals. How would an AV react to a police officer directing traffic? Moreover, considering how AV technology is fed and improved as more driving behavior data is collected, I think of the many distinct driving behaviors. Someone in New York City will not drive the same as a person in Minneapolis or St. Louis, for example. A teenager will drive differently than a seventy-year-old, and so forth.

Also speaking to the problem of having state-based AV regulations, Paula Bejarano, autonomous vehicles technology scout at Goodyear, considers that for specific use cases like ridesharing it may not be a huge impact as long as AVs are staying within city limits. However, an autonomous shuttle across state lines to a nearby airport, or an autonomous truck driving across multiple states along the I-10 highway corridor, may face local restrictions.

This resonated with me. Living in Washington, DC and driving very often through the Washington Metropolitan Area, which includes surrounding areas of Virginia and Maryland, would certainly be a challenge. Likewise, to Bejarano's airport example, for those of us in the District of Columbia, the three nearby airports—Ronald Reagan DCA, Washington Dulles IAD, and Baltimore/Washington BWI—are all outside the perimeter of the city. This would result in the hassle of having to change license plates, even when in a rush not to miss our flights. As if traffic congestion wasn't enough of an inconvenience.

Bejarano assures there is a handful of states encouraging the deployment of AVs much more than others, which can put more barriers to their entries. Interestingly enough, Bejarano says although California is undoubtedly the center of AVs development, it's a state seeing numerous regulatory roadblocks.

"You see Arizona being very friendly toward autonomous vehicle testing, Texas, as well as Vegas, and this is where many companies are deploying their prototypes or their proof of concept because there is no kind of regulation. So, it's like having a carte blanche; you can freely do road testing of autonomous vehicles. Then there are a lot of states with autonomous driving as a way to bring in business, so they're more friendly and they're more welcoming."

"Funny enough, in California the hub of vehicle development, there are a lot of barriers to deploy autonomous vehicles commercially since a lot of companies are not allowed to make revenue. I think it's because they look at it like you're testing the technology, so you shouldn't test with human customers inside."

Hearing Bejarano speaking to the issue of regulation in autonomous driving, I remembered I had come across a *Mobility* podcast interview of David Estrada, Nuro's chief legal and policy officer. Nuro is an American robotics company known for their self-driving vehicles designed for local goods transportation. During the interview, Estrada discussed the origins of the company and how he helped Nuro navigate the NHTSA's approval of the first-ever exemption process of a fully autonomous vehicle from three specific Federal Motor Vehicle Safety Standards (FMVSS) that were clearly unnecessary for a driverless vehicle carrying goods.[99]

Nuro was founded in 2016 by Dave Ferguson and Jiajun Zhu, two of the people who were in the project formerly known as Chauffeur at Google which turned into Waymo. Waymo is an American autonomous driving technology development company and subsidiary of Alphabet, Inc., the parent company of Google.

Estrada claims he was at a Chauffeur program at the same time as Nuro's founders back in 2011. He explains how Google was the first company to really get interested in AVs, assembling a team of people to design self-driving vehicles in the United States. Estrada says they built the Google X cloud, which is the moment he became involved as a lawyer, advocating for the creation of laws favoring AVs deployment. He would after play a role in the creation of the first self-driving car video.

"I was a lawyer in another part of YouTube and Sebastian Thrun [former Google VP] had pulled together this team that wanted to put a product out, and they wanted this product to

---

99 David Estrada, "#065: David Estrada, Nuro," February 18, 2020, in *Mobility Podcast*, produced by Greg Rogers, podcast, Apple Podcasts, 1:01:13.

drive itself on public roads. But they asked themselves, 'can this be done legally?' And so, they needed a lawyer to come help and I joined in and helped the team pass the first laws in the United States for self-driving cars, and those were in California and Florida."

"And those of us who had kept on with the program, he called us on this one project which was to create a video that would show, for the first time, the self-driving car. I worked on that project, and so did Nuro's founders Dave Ferguson and Jia-jun Zhu, and what we showed was a gentleman named Steve Bannon going out and performing a drive."

"And how this relates to Nuro is the car picked him up at his home and it took him to run some errands. It took him to Taco Bell because he really liked Taco Bell, and it took him to pick up the dry cleaning and took him back home."

Estrada says this experience captured Nuro's stakeholders' imagination of what self-driving cars could do. It makes reference to how today most companies are still focused on the idea that self-driving cars should pick people up or take them places. On the other hand, Nuro is taking a very different approach upon realizing most of what people do when they get in their cars is go run errands.

The question that came to my mind was should we put human beings in self driving cars or should the car itself go to pick things up and bring them to us?

For this use case at least, Nuro has nailed it down by turning into a car for delivery of products, many of which come from internet shopping but also from local delivery of hot food services and grocery delivery services. This way, Nuro is helping people to use their time in different ways as opposed to always running out and conducting errands.

Further, I thought of the numerous occasions during my time working as a structural design engineer and bridge inspector where I would drive into Maryland rural areas to inspect bridges and culverts. Aside of just a couple spread out convenience stores, I did not see grocery stores for miles, and always wondered where people living in those so-called food deserts would purchase their goods. More so, I thought of the long vehicle drives and associated carbon emissions required for access to these.

Then I think about the benefits for people living in those rural communities to have their products delivered to their home from a store two to three miles away. Not only would they avoid a four-to-six-mile roundtrip drive, but they also save money on their purchase of products which we know are (nearly) always more expensive at convenience stores. More so, Estrada explains delivery is free.

"We can really do something very important for these food deserts, some really important societal change."

Given the atypical structure of Nuro's vehicles without the need or room for a passenger, Estrada questioned the necessity to satisfy all FMVSS. On behalf of the company, he successfully received the NHTSA's approval of the first-ever exemption process of a fully autonomous vehicle from the inclusion of windshields, mirrors, and rear-facing cameras, which is only relevant if a passenger is inside. According to Estrada, there are roughly eighteen more standards (e.g., airbags, seat belt harness) which are also unnecessary because Nuro's vehicles are already safe by being smaller, narrower, and lighter than an economy car.

COLLABORATION TOWARD AUTONOMY

There is a whole stakeholder ecosystem working toward the successful implementation of AVs. Startups are playing a key role in the autonomous space. They are bringing immense value to the advancements in AVs, and an advantage original equipment manufacturers (OEMs) don't have. They can focus on one specific area rather than trying to look at the whole package.

Bejarano suggested during our interview that no company, large or small, could deploy an AV by itself. As such, she provided the example of ridesharing company Uber, and how their self-driving program hasn't evolved fast enough because they don't have the capital resources, and their margins are slim. Then you have the OEMs, one-hundred-year-old companies, with established engineering procedures and typical development cycles ranging from three to five years. They've mastered the hardware assembly of the vehicle but struggle with the fast innovation on the software side.

Bejarano explains although startups have the ingenuity and a super-fast and very agile development process, they are very limited in resources as they don't have the hundreds of millions of dollars that larger companies do. In fact, startups don't have the footprint, meaning they can only scale at a small lens such as in a specific neighborhood or city. This is where the smaller players need and should leverage the nationwide presence of transportation network companies (TNCs), OEMs, and Tier-1 suppliers, among other players.

Bejarano suggests General Motors (GM) and Ford, amongst other automakers, are realizing the new entrants, such as electric vehicle company Rivian, are offering their customers a new experience. Hence, OEMs will not own the entire mobility ecosystem and Bejarano underscores the necessity for a collaborative environment toward the successful deployment of AV technology.

"Today competitors are working together. They realize they can't do it alone, so it will become a joint effort. Every player will contribute a small part of the AV stack algorithms or an improved hardware integration, and eventually in five to ten years, the challenges encountered today will be overcome. Eventually, we will reach an efficient scenario where technologies such as the software that drives the vehicle will become a commodity."

I could not agree more; as mentioned earlier, automakers are starting to reshape their roles as more than vehicle manufacturing companies. They have multiplied their efforts to invest, partner with, acquire, or create mobility subsidiaries.

The next puzzle to solve, Bejarano says, will be the business models around automated driving. For this area, peripheral stakeholders will get involved, such as retail organizations like the Walmarts and Amazons of the world. Similarly, AV integrators will have to work with transportation departments and insurance agencies, among others. So, it's really an ecosystem of collaboration, Bejarano manifests.

To put further emphasis on the AV collaborative ecosystem landscape, Bejarano shared with me some pieces of the work she comes across at her job at Goodyear. More specifically, she touches upon the fact that nowadays, software and data are at the center of the transportation industry. As such, Bejarano tells her journey in collaboration with startups.

"My job is to figure out, okay, there is startup A that could solve this problem we have in-house. So, how to bring them together and how to connect them to the right people. Because the industry is moving away from a hardware focused engineered model to a software focus."

"So, today everything is connected, everything is about data. It's like every component of a vehicle would be connected so I also do that on a daily basis. I'm always looking at ways to send data faster, or simplify this kind of data we move from point A to point B. How does the vehicle talk to the infrastructure? How does the vehicle text another vehicle? How do we predict things on the vehicle?"

There are indeed established stakeholder ecosystems and a plethora of players collaborating in the AV space toward the full implementation of self-driving vehicles. Automakers and auto-parts providers including Ford, Volkswagen, GM, Tesla, Volvo, Nissan, and Toyota. Technology providers include Intel, Delphi, Apple, and Google. Services providers Tesloop, Lyft, Uber, and RydeCell, and self-driving startups Aurora, TuSimple, and Otto are only examples of an extensive list of mobility companies putting efforts into making AVs a true reality.[100]

Moreover, in the case of Ford, for example, in July of 2019 the company purchased Quantum Signal, which is behind computer-generated environments used by militaries to test unmanned remote and autonomous systems.[101] Ford has also partnered or invested with four different technology companies, doubling its presence in Silicon Valley. Similarly, in 2019 Volkswagen Group announced an investment of $2.6 billion in Argo AI, the autonomous vehicle startup based in Pittsburgh, and made a significant investment in Aeva, a startup research on vision sensors for driverless cars.[102]

MORE THAN AN AUTOMATED PRIVATE PERSONAL CAR

While the concept of an autonomous vehicle appears to invoke an upgraded replacement of the traditional private personal car, autonomy technology is also being leveraged for other use cases, as already seen with the case of Nuro's product delivery vehicle. In the ride-hailing ridesharing services arena, in June of 2019, Uber acquired Seattle startup Mighty AI's intellectual property, tooling, tech talent, and labeling community.[103] Similarly, Lyft partnered with Waymo in May of 2019 to bring self-driving vehicles onto the ride-hailing network in Phoenix as the company ramped up its commercial robo-taxi service. Likewise, and aside from the efforts

---

100 "Top 30 Self Driving Technology and Car Companies," GreyB, accessed October 23, 2020.
101 Ibid.
102 Ibid.
103 Ibid.

in drone delivery, in 2020 Amazon acquired Silicon Valley self-driving startup Zoox and unveiled a self-driving car which could be the e-commerce giants first robo-taxi.[104]

Automated technology is being implemented across other vehicle types and applications such as long-haul trucking. For instance, a self-driving truck company called TuSimple is capable of driving from depot to depot without human intervention needed and is offering its trucks for purchase to fleet managers via their OEM partners. Paula Bejarano explains in a blog article how, except for OEMs such as Tesla, Volvo, and Daimler, not many other players in the industry are taking the vertically integrated approach for self-driving trucks.[105]

In 2019, TuSimple completed an autonomous truck test for the United States Postal Service (USPS) delivering letters and packages between distribution centers more than one thousand miles apart, without incidents whatsoever.[106] Retail corporation Walmart partnered with software company Gatik to deploy AV pilot programs to move customer orders from a fulfillment center to a market in Arkansas.[107] I think of how, similar to USPS, private delivery services companies

---

104 Sean O'Kane, "Zoox Unveils A Self-Driving Car That Could Become Amazons First Robotaxi," *The Verge*, December 14, 2020.

105 Paula A. Bejarano, "Robo-Trucks: Could they be first?," *Medium*, July 21, 2019.

106 Murray Slovick, "TuSimple Completes Self-Driving Truck Test for the USPS," Electronic Design, June 24, 2019.

107 Tom Ward, "Walmart and Gatik Go Driverless in Arkansas and Expand Self-Driving Car Pilot to a Second Location," Walmart, December 15, 2020.

FedEX and UPS could substitute their vans and trucks with autonomous fleets.

In sync with the idea of a diversified autonomous vehicle portfolio and considering AVs will become very expensive, Bejarano considers the future of AVs will be a fleet model one rather than an old model of vehicles. Hence, she believes AVs will only be owned by fleet operators to serve individuals in our societies.

"So there are so many new startups that are using very different business models to take away that car ownership which is where we are heading. You'll see the OEM launching those new initiatives of having fleets and being able to provide them to the people, rather than selling to them."

In my opinion, while AV usage would be beneficial for all segments (e.g., private car, ride-hailing), a model where public transportation agencies or private companies leverage their fleet (e.g., buses, shuttles) to supply AV services to all people is particularly desirable. As for the benefits of autonomous long-haul freight, other than the inherent convenience, I think of the numerous truck accidents on US highways—resulting in 46,000 injured people and 892 deaths in 2019.[108] A well-implemented system of fully autonomous trucks would certainly reduce the probability of unfortunate human and economic losses.

---

108 *Overview of Motor Vehicle Crashes in 2019*, National Center for Statistics and Analysis, Traffic Safety Facts Research Note, Report No. DOT HS 813 060 (Washington DC: National Highway Traffic Safety Administration, 2020).

ACCOMMODATING ALL RIDERS

As I mentioned at the beginning of this chapter, fully automated vehicles (AVs) have, or should have, a broader mission than providing seamless convenient transportation for those who prefer not to drive. Aside from representing a solution to increasing traffic congestion and vehicular accidents, policy makers, private mobility companies, and public agencies will be urged to work together to ensure AVs are reaching and effectively serving all the population.

According to data from the US Department of Transportation, in 2018 an estimated 25.5 million people in the US had disabilities that made travelling outside home difficult.[109] We all know how transportation unavailability affects humans' lives. It prevents them from accessing jobs, resources, and other social privileges.

A promising sign, however, is we are seeing some companies putting efforts into having driverless technology be a solution for customers with mobility, vision, and hearing impairments, including seniors and those with chronic health conditions. Alphabet's Waymo, for example, is engaging in discussions with the Foundation for Senior Living in Phoenix and the Foundation for Blind Children.[110] Similarly, Cruise has partnered with the American Council of the Blind, the National Federation of the Blind, and local communities to conduct usability studies and solicit feedback.

---

109 Stephen Brumbaugh, "Travel Patterns of American Adults with Disabilities," US Department of Transportation, last modified December 11, 2018.

110 Kyle Wiggers, "Autonomous Vehicles Should Benefit Those with Disabilities, but Progress Remains Slow," VentureBeat, August 21, 2020.

Other examples of initiatives to promote the access of AVs services to people with disabilities include Local Motors 2018 launched shuttle that directed visually impaired passengers to empty seats using machine vision.[111] Likewise, May Mobility developed a secure wheelchair-accessible prototype version of its AV. In May 2019, Volkswagen (VW) launched their Inclusive Mobility Initiative (IMI) as an effort to democratize mobility. Through this initiative, VW is working with community groups and disability advocacy entities such as the Disability Rights Education and Defense Fund and the Light House for the Blind and Visually Impaired.

THE FUTURE IS AUTONOMOUS

Despite the technological and regulatory challenges involved with the implementation of autonomous vehicles, recent advancements in self-driving technology hold the premise of fully automating the transportation industry over the next few decades. The United States is well positioned to continue making advancements in this space.

During a talk at the Uber Elevate Summit in Washington, DC in June of 2019, the US Department of Transportation Secretary claimed more than 1,400 self-driving vehicles were in testing then by over eighty companies across the United States.[112] According to 2019 reports submitted to the California Department of Motor Vehicles, autonomous vehicle companies claimed their vehicles drove nearly 2.9 million miles in the US during the most recent reporting period. This

---

111 Ibid.
112 Darrell Etherington, "Over 1,400 Self-Driving Vehicles Are Now In Testing By 80+ Companies Across The US," TechCrunch, June 11, 2019.

was for AV firms with a permit to test their technologies with a safety driver only and represented an increase of more than eight hundred thousand miles from the previous reporting cycle.[113] The Brookings Institute estimates the investment in autonomous vehicle technology over the past three years to be $80 billion across more than one hundred sixty investments, partnerships, and acquisitions.[114]

Considering the abovementioned facts in developing self-driving technology, in addition to the notorious AV advancements, it should not surprise us the AV market is expected to exponentially grow.

The global autonomous vehicle market size valued at $54.23 billion in 2019 is projected to reach $557 billion by 2026, expanding at a compound annual growth rate (CAGR) of 39.5 percent from 2019 until 2026.[115] To put this in perspective, the CAGR for the global healthcare industry is *only* 13.4 percent from 2019 to 2025.[116]

I am positive the autonomous vehicle industry will continue transforming the US transportation sector over the next few decades, decreasing ongoing vehicular traffic accidents,

113 "AV Permit Holders Report Nearly 2.9 Million Test Miles in California," State of California Department of Motor Vehicles, Office of Public Affairs, February 26, 2020.

114 Cameron F. Kerry et al. "Gauging Investment In Self-Driving Cars," The Brookings Institution, October 16, 2017.

115 "Autonomous Vehicle Market Outlook - 2026," Allied Market Research, accessed January 15, 2020.

116 "Healthcare CRM Market Size to Exceed USD 21.46 Billion by 2025, at 13.4% CAGR, Says Market Research Future (MRFR)," Market Research Future, GlobeNewswire, February 8, 2021.

greatly reduce traffic congestion, and enhance transportation availability for all.

KEY TAKEAWAYS

- Advanced driver-assistance systems (ADAS) enabled by the great progress on hardware technologies (e.g., radar, LiDAR), as well as the evolution of AI, big data, and IoT show promise to contribute to a reduction in vehicular traffic accidents in the United States.

- Although there are many companies working toward the deployment of fully autonomous vehicles, there are still many technological and regulatory challenges to overcome. Ensuring self-driving vehicles are safe and making AV regulations federal-based rather than state-based, are not easy tasks.

- The implementation of AVs is being pursued by an entire collaborative stakeholder ecosystem comprised of OEMs, Tier-1 suppliers, mobility services companies, and technology providers startups.

- While efforts in AVs have mostly centered around the traditional private passenger car, self-driving technology is gradually being leveraged for other use cases and vehicle types, including product delivery, ridesharing, and long-haul trucking. Likewise, the future of AVs calls for an inclusive fleet-based model to be owned by fleet operators, that can drive effective mobility for all the US population.

- Policy makers, transportation agencies and mobility companies must ensure AVs are deployed in a way they can serve all peoples transport needs, including individuals with disabilities.

CHAPTER 6:

MICROMOBILITY

"Life is like riding a bicycle. To keep your balance, you must keep moving."[117]

—ALBERT EINSTEIN

Chances are in your early childhood you learned how to ride one. Whether you are a cycling aficionado or just have one sitting in your garage, you can probably at least keep your balance while pedaling one. I'm talking about the bicycle. They're *cool*, they're convenient, and you only need to learn how to ride it once. As the popular saying goes "you never forget how to ride a bike".

I remember when I got my first bike from my parents sometime in the mid '90s. I would use it mainly for fun but occasionally also used it for transportation purposes, such as mini adventures in my neighborhood in suburban Quito Ecuador or trips to the grocery store to satisfy my mother's mandates. That bike took me dozens of miles traveled at summer camps

---

117 Rex Hammock, "Einstein Explains Why Life is Like Riding a Bicycle," Motivation, SmallBusiness, November 4, 2016.

through the middle of the Ecuadorian Amazon Rainforest to downhill mountain biking races which, by the way, I was awful at.

The popularity of traditional recreational bicycles has remained fairly constant even two decades later, however a new concept around bikes has emerged over the past years—bikeshare. No, I'm not referring to tandem bicycles designed to be ridden by more than one person. I'm describing services in which bicycles are made available for shared use to individuals on a short-term basis, for a price or free on certain occasions.

The Institute for Transportation and Development Policy (ITDP) defines micromobility as a range of small, lightweight devices typically not exceeding a speed of 15 mph (25 km/h) and ideal for trips up to six miles.[118] It includes privately owned or shared, human and electric modes like bikes and scooters—docked and dockless—but excludes combustion engine vehicles and those moving faster than 28 mph (45 km/h).

The trend of micromobility has gained impressive momentum as it includes a variety of more sustainable transportation modes over private automobiles. Due to its convenience and affordability, micromobility has been considered by many mobility experts as the future of urban transportation,

---

118 "Defining Micromobility," Multimedia, Institute for Transportation & Development Policy (ITDP), accessed September 22, 2020.

especially in light of first- and last-mile transport.[119,120] The "first- and last-mile" term describes the beginning and end of a person's public transport journey.

More broadly, micromobility is perceived as an effective transport method at "transit deserts," which are areas with transit-dependent populations that lack adequate public transit service. As such, micromobility helps bridge the gap between the level of transit services and the transport needs of communities.

According to data from the National Association of City Transportation Officials (NACTO), in 2019 there were one hundred thirty-six million trips on shared bikes and scooters, a 60 percent increase from 2018.[121] This number came from forty million trips on station-based bikeshare systems (pedal and e-bikes), ten million trips on dockless e-bikes, and eighty-six million trips on scooters. In 2019, one hundred nine cities had dockless scooter programs, a 45 percent increase from 2018. NACTOs report indicates this contributed to an over 100 percent increase in trips taken on scooters nationwide. Scooters indeed appeared to be more than glorified kids' toys.

So how exactly do micromobility services operate? It all starts with a mobile phone and a downloadable app. In fact,

119 Nitin Lahoti, "Micromobility: The Next Wave Of Eco-Friendly Transportation," *Blog*, Mobisoft Infotech, October 14, 2019.

120 Rasheq Zarif et al., "Small is Beautiful," Deloitte Insights, Deloitte, April 15, 2019.

121 *Shared Micromobility in the US: 2019* (New York, NY: National Association of City Transportation Officials (NACTO), 2020).

all these fascinating and innovative mobility options are emerging from the explosion of data and technology. The digital revolution!

Moving on to the riding sign-up process, while it can differ across distinct bike companies (e.g., Citi Bike, Capital Bikeshare) or scooter firms (e.g., Bird, Lime), it all begins with the user entering her or his credit card details in the app. Next, they select a few scooter options and upload a photo of their driver's license. Users then look up nearby scooters in the apps map, unlocking the available vehicle by scanning the products bar code, and simply parking the scooter in safe designated areas once done. In the case of docked bikes, they return it to a dock station. Pricing options differ between scooters and bikes, with the former priced per minute and the latter priced per mile.[122,123]

DOCKED BIKESHARING: AN EFFICIENT AND SUSTAINABLE PUBLIC-PRIVATE EFFORT

Laura Fox is Lyft's general manager and director for Citi Bike, a privately owned public bicycle-sharing system in New York.[124] During a *Micromobility* podcast interview, Fox shares her experience as a cyclist in New York City, and provides insights on the Citi Bike setup, docked systems, and its associated advantages.[125]

122 Ethan May, "Here's Everything You Need to Know About Bird and Lime Electric Scooters," *IndyStar,* last modified September 25, 2019.

123 "Single Trip," Capital Bikeshare, accessed September 23, 2020.

124 "Citi Bike," NYC Bike Maps, accessed September 23, 2020.

125 Laura Fox, "82: The Biggest Bikeshare In America—Talking with Laura Fox, Lyft's General Manager for Citi Bike in New York," July 23, 2020, in *Micromobility Podcast,* produced by Oliver Bruce, podcast, 1:07:03.

"I have been a Citi Bike rider since I moved to New York in 2013, and I found it to be a transformational way to ride around the city. And so, it brought me to thinking about how I take all these various ideas I've been thinking about and working overtime and bring them into a really practical program that can dramatically improve someone's quality life in terms of getting to work."

"Thinking about sustainability metrics, in terms of load shift, until it's kind of a real living business and a great public-private partnership for the city of New York."

Citi Bike launched in 2013 and the company now has over one thousand stations and over fifteen thousand bikes. Fox explains how Citi Bike has a mandate with the City of New York and Lyft to embark on a $100 million expansion by 2024, where the company will be doubling the size of their service area. As a matter of fact, Citi Bike's history was grounded in a strong, long-term, and reliable public-private partnership. The idea came from the Department of Transportation (DOT) with the intention of bringing a bikesharing model to New York similar to the one in London (UK). In the hopes of making it sustainable, the DOT financed and assigned the operations of Citi Bike to the private sector.

"It's a long-term, reliable partnership, and our partnership with the DOT is central to the role Citi Bike can play in helping people get around. And I guess, saying it another way, Citi Bike wouldn't be as successful as it is today without New York City and the DOT being so dedicated."

This is great evidence of how public policy and public transport agencies can enhance mobility in our cities.

Moreover, Fox suggests the Citi Bike initiative has more than a simple private business goal and speaks to New York City's policy objective. She describes Citi Bike's membership option for riders as an effective method to help reduce the number of miles driven in cars and other less sustainable modes of transportation. A Citi Bike member will on average take fifty-five to sixty bike rides per year, compared to only 2.5 rides per year for the casual rider. The membership option is thus very attractive since Citi Bike member riders will pay less than if buying a subway pass for two months.

Dock setups for the bikes bring advantages. They create a reliable network and infrastructure system across the city of New York which, in turn, provides a functional connection into neighborhoods, allowing people to get to places faster. Dock systems provide an effective use of public space and, more specifically sidewalk access, avoiding street clutters in crowded places like Manhattan.

"So you could have this kind of orderly distribution that avoids street clutter, minimal interference with right-of-way and sidewalks. So, thinking about accessibility concerns and things like that."

Additionally, dock setups contribute to the original mission of Citi Bike centered on shifting modes of transportation. Fox suggests the dominant peak times for their bike product are in the morning and afternoon, corresponding to commuting patterns and representing about 80 percent of ridership. Dockless bikes, on the other hand, have higher ride volumes on weekends and see their peaks during lunchtime on weekdays.

"And all of those [dockless bikes] can have really great use cases, but I think with dock products you can really see this big move toward solving how to get people to use it for that longest distance trip, probably throughout their day and their commute, as a simple node."

As I was listening to Fox speak to the city of New York and Citi Bike public-private partnership for the bikesharing system provided by Lyft, I remembered I had come across a very similar type of partnership in Portland, Oregon. I'm referring to a July 2020 announcement by the Portland Bureau of Transportation (PBOT) and Lyft on the extension and expansion of BIKETOWN, Portland's bikeshare system.[126] Interestingly enough, sports apparel company Nike, Inc. is the founding partner for this initiative starting in 2016 which, in its first year alone, saw more than seventy-five thousand riders and over six hundred thousand miles biked. Further, the new contract proposed the deployment of an all-new fleet of 1,500 electric pedal-assist bikes.[127]

It's another piece of proof of how private-public partnerships can bolster the transportation ecosystem for our communities.

---

126 "PBOT Announces New Biketown Agreement With Lyft And An Extension of Its Title Sponsorship With Founding Partner Nike, Inc. for Portland Bike-Share Through 2025," City of Portland, Oregon, July 16, 2020.

127 Ibid.

I have personally experienced Washington, DC's bikesharing system Capital Bikeshare since 2012. Docked bikes have, in fact, represented a smart and effective way for me to avoid the traffic congestion and parking difficulties I would deal with if driving a car. Moreover, it's an affordable transport option and unless you are dealing with extreme weather conditions, biking is also a fun and healthy activity. Everyone can do it regardless of past experience. Capital Bikeshare has deployed effective infrastructure network of docks across all eight DC wards, providing accessible transportation resources for everyone.[128]

THE RISE OF THE ELECTRIC SCOOTER

While Lyft is a synonym for ridesharing, the story shared by Laura Fox clearly speaks to the company's success and strong presence in bikesharing services. Additionally, Lyft has deployed electric scooters across US cities, another micromobility option emerging in the past few years. Moreover, use cases for Lyft micromobility options include serving last mile logistics companies such as Grubhub food delivery service.[129] Similar to Lyft, there are a number of other firms putting efforts into the deployment of electric scooters in the United States.

This is the case of Bird, the first electric scooter company to launch in the US, having first deployed in September 2017

---

128 "East of the Anacostia River Network Expansion," Capital Bikeshare, accessed September 25, 2020.

129 "Lyft and Grubhub Team Up to Bring Unlimited Free Delivery From Your Favorite Restaurants and Other New Perks to Lyft Pink," *Blog*, Lyft, October 6, 2020.

in Santa Monica, California.[130] During an interview at the 2019 Upfront Summit—a Los Angeles premier technology event—CEO and founder of Bird, Travis VanderZandern discusses what inspired him to start the company.[131] He also refers to the substantial growth and rapid adoption by riders, as well as challenges the company faces.

"My mother was a public bus driver for thirty years and so I grew up passionate about transportation and seeing it firsthand every day... what I love and I'm passionate about doing is using technology to move people around the world."

VanderZandern ended up on the West Coast and spent over four years in the ridesharing space between Lyft and Uber, during which he witnessed a very positive impact of ridesharing on transportation. More specifically, in reducing DUIs and alleviating parking difficulties. Nonetheless, what really made him reflect was the high number of short distance ridesharing trips, triggering his motivation to start Bird.

"But one of the things I realized in my time in ridesharing was we were using four-thousand-pound vehicles to move people around mostly less than five miles. It's incredible how short distance your trips actually are. And so, I left Uber and started incubating Bird."

130 Megan Rose Dickey, "The Electric Scooter Wars of 2018," TechCrunch, December 23, 2018.

131 *Upfront Ventures,* "Travis VanderZanden Interviewed by Mark Suster | Upfront Summit 2019," March 25, 2019, video, 25:18.

VanderZandern explains how when he left Uber, what he really wanted to do was last-mile dockless electric transportation. He started importing electric bikes, electric skateboards, among other electric vehicle types from China. This was around the same time he bought bikes for his two daughters for Christmas. Although they enjoyed riding the bikes, they asked him for their scooters again. This would represent a crucial moment for VanderZandern at realizing how scooters were a preferred electric mode of transportation.

"They [VanderZandern's daughters] loved the bikes and they rode them all day long and they got excited, but then the next day they woke up and they said, 'Daddy, can we ride our scooters again?' And it was kind of an 'aha' moment for me at a very fortunate time because I had been thinking about what is the form factor that makes most sense for short-range electric transportation for adults."

"So I bought an electric scooter, and I would ride it around. People would always ask me about it, and the people who rode it would get addicted and wanted to keep riding. So, at that point, I got pretty excited to actually launch Bird with electric scooters."

According to Bird's founder, however, a lot of people were very skeptical at first of electric scooters until they actually rode them. Then they fundamentally saw how they could change transportation in a big way. Individuals indeed realized by being willing to get out of their cars and use another mode of transportation, it ultimately could have a big impact on the world.

I can surely testify to VanderZandern's point on the initial skepticism scooters generated. As someone who used Bird and other electric scooter options in numerous occasions in Washington, DC, I have to admit they didn't seem, at first, to be the safest transport option. Moreover, the idea of leaving the scooter parked almost anywhere after using it was questionable in the early days. What if it gets stolen or damaged? Will Mr. Bird hold me accountable and come after me? Those were a few thoughts that crossed my mind back in the day. Nowadays, electric scooters are increasingly being perceived to bring immense transport benefits, especially for short trips.

Bird rapidly expanded in less than a year from one city (Santa Monica) in 2017 to over one hundred cities across the US, Europe, and Mexico City, in addition to one hundred college campuses. Cities have quickly opened their arms and allowed electric scooters to operate in numerous cities. In VanderZandern's view, this is due to the company's mission of reducing traffic and carbon emissions, which is in line with most cities' goals.

Yet, he explains how the deployment of Bird e-scooters in Santa Monica was a little rocky as the city hadn't thought about electric scooter sharing when creating regulations. Fortunately for Bird, by working together with city's stakeholders, Santa Monica has fully embraced electric scooters today. There is now, in fact, spray painting dedicated parking on the sidewalks throughout the city to allow the scooters to park.

This is in my opinion particularly important. You don't want scooters to become a hazard for pedestrians if left anywhere on a sidewalk. In fact, when it comes to transportation and business, safety has to be the top priority.

To this point, VanderZandern claims since day one, Bird has tried to prioritize safety over growth by, for instance, not allowing riders to ride scooters after 9:00 p.m. The company has actually seen across the industry an increase in accidents and deaths on some of the other platforms after midnight. Thus, although Bird could get more rides after midnight—or 9:00 p.m. which is their current philosophy—they think it's not worth risking people's lives over squeezing out a few extra rides.

Moreover, VanderZandern discusses how when they first launched in Santa Monica they didn't require a driver's license when a rider signed up, and yet they enforce it now. He considers this initiative to be very important given that safety comes before profit.

"And early on you would see ten- or twelve-year-old kids riding the scooters, and as a father of two young daughters it terrified me. So, we quickly required a driver's license on sign up. We would verify you were eighteen or older and we're hoping the rest of the country follows suit on that. We think there are certain things like safety that have to be prioritized over growth, and we think it's the right thing to do."

While I strongly agree with the statement that safety comes first, I believe a driver's license enforcement eliminates a portion of the population which might still be able to ride scooters safely. More so, I think of sixteen-year-olds who are just getting their first motor vehicle driver's license. These teenagers could find in e-scooters a more attractive, effective, and even safer transport alternative to driving a personal car for those shorter trips.

In addition, VanderZandern refers to the challenge of increasing the number of bike lanes in cities due to local people protesting against city governments anytime these have been added. Interestingly enough, VanderZandern affirms local politicians have for the first time admitted they could begin to change the way cities are laid out due to the increasing demand scooters are seeing.

Along the same lines, VanderZandern admits when he first started Bird and was thinking about what group would be against scooters—similar to taxis against ridesharing—he believed bikers would possibly oppose the deployment of scooters. It turned out bikers loved scooters because they're drawing attention to the fact there are not enough bike lanes and not enough protected bike lanes. As such, Bird is trying to work with cities, even financially, to help address this issue. This has already been tackled in Europe through the deployment of protected bike lanes, thus making it safer to ride a scooter.

"The cities have always wanted to create more bike lanes, but they've always been in this chicken and egg problem where if they take away a lane from traffic or parking, people will be angry."

"They didn't think Americans would actually get out of cars and use bike lanes, and we're showing with Bird and other transportation companies that Americans will get out of their cars."

This is certainly a hot topic which has brought lots of tension in mobility among urban planners and transportation engineers. I personally believe building more dedicated bike/scooter lanes is providing more freedom of choice for transport users. The more transportation types we can offer to individuals, the better. It's also important to recall micromobility trips do not replace all car trips. They appear as an effective, sustainable, and affordable solution for first- and last-mile transportation. Let's give people a menu of transport choices and the associated infrastructure so they can optimize their trips based on real times and conditions. Let's incentivize them to adopt more ecofriendly transport services.

Lastly, VanderZandern talks about the success Bird has had on college campuses, and how the company is putting efforts into providing the safest possible experience. For example, by automatically reducing the scooter's speed when entering areas the college may consider slower speed zones. So, unlike cars, with Bird you can enter a *geohash* area and the scooter will slow down from 15 mph down to 10 or 5 mph. This is, in fact, already being done on the Santa Monica Beach bike path and around elementary schools.

"Imagine if cars, when driving by an elementary school during school hours, actually automatically slowed down. That's what's happening already with scooters and it's pretty exciting."

In short, VanderZandern defines Bird's mission as making cities more livable, meaning less cars and less carbon emissions. Bird, among other scooter companies, are increasingly penetrating cities in the US and providing people an alternate effective method of transportation. VanderZandern excitingly reveals these services have seen a rapid and growing adoption, particularly with shorter trips.

ENSURING EQUAL ACCESS TO MICROMOBILITY SERVICES

Micromobility services like docked bikes and e-scooters have certainly demonstrated success in US cities over the past few years. They have begun enabling easy and effective first- and last-mile transportation access to the population. It is important, however, to ensure new mobility services are available to all residents, regardless of where they live.

In a report published in the National Institute for Transportation and Communities (NITC), professors Rebecca Lewis, PhD, and Rebecca Steckler from the University of Oregon share their views on the impact of emerging technologies and new mobility on cities. They emphasize the benefit of reducing vehicle miles traveled and congestion through a shift in modes of transportation. Yet, they point out the issue of transportation inequality rises as an implication of this transport change.[132]

---

132 Rebecca Lewis et al., *Emerging Technologies and Cities: Assessing the Impacts of New Mobility on Cities*, Report No. NITC-RR-1249 (Portland, OR: Transportation Research and Education Center (TREC), 2020).

To mitigate this disparity in transportation access, an increasing number of cities are requiring mobility companies to disperse their vehicle solutions (e.g., bike, e-scooter) in low-income or underserved neighborhoods or across the entire city.

In Washington, DC, for example, it's a requirement to have e-scooters in every ward. Some jurisdictions require companies to offer diverse payment options considering low-income people tend not to have a smartphone or credit card, and thus cannot put a deposit down. The study report notes, in fact, DC requires companies to offer unlimited thirty-minute trips to customers who are at the 200 percent federal poverty level.[133]

Similarly, and with the intentions of bridging the gap in transportation in US cities, the San Francisco Municipal Transportation Agency (SFMTA) launched a pilot program in early 2020 to test adaptive scooters for people with disabilities.[134]

Overall, professors Lewis and Steckler refer to Seattle and Washington, DC as two great examples of cities incorporating equity into new mobility strategies and policies:

"Seattle's new mobility playbook includes principles, plays, and actions to improve safety... The strategies in this play include

133 Rebecca Lewis et al., *Emerging Technologies and Cities: Assessing the Impacts of New Mobility on Cities*, Report No. NITC-RR-1249 (Portland, OR: Transportation Research and Education Center (TREC), 2020).
134 "SFMTA Launches Pilot Program to Test Adaptive Scooters for People with Disabilities," *Mass Transit*, January 22, 2020.

enhancing transportation services for vulnerable groups such as the LGBTQIA+ community, youth, seniors, people with disabilities, and many others to ensure everyone can access smart phone services, ensure a wide array of payment options, make sure new mobility services are ADA accessible, and more."[135]

"The District adopted a new e-scooter and motorized bicycle permit (effective January 1, 2019) that requires e-scooter coverage in every ward (eight total) and allows up to six hundred e-scooters per company with the potential to increase that amount by 25 percent every three months. In addition, companies are encouraged to offer adaptive vehicles that can accommodate people with mobility devices (like wheelchairs). These vehicles are not counted toward the total allowed."[136]

THE FUTURE

The ongoing deployments of bikesharing systems and electric scooters, in addition to increasingly favorable city regulations, are setting micromobility as a transportation industry disruptor.

The micromobility market size is already massive. According to a McKinsey & Company study, micromobility could encompass all passenger trips of less than eight kilometers

---

135 *New Mobility Playbook*, City of Seattle (Seattle, WA: Seattle Department of Transportation (SDOT), 2017).

136 *DDOT Releases New Permit Application for Dockless Vehicles*, Government (Washington, DC: District Department of Transportation (DDOT), 2018), quoted in Rebecca Lewis et al., *Emerging Technologies and Cities: Assessing the Impacts of New Mobility on Cities*, Report No. NITC-RR-1249 (Portland, OR: Transportation Research and Education Center (TREC), 2020).

(five miles), which represent 50 to 60 percent of today's total passenger miles traveled in the United States.[137] For instance, roughly 60 percent of car trips are less than five miles and could use micromobility solutions, which could also substitute roughly 20 percent of public transportation travel, in addition to trips conducted in private bicycle, scooter, or simply walking.

McKinsey's forecast model disclosed a 2030 market potential of $200 billion to $300 billion in the United States.

I am confident sustainable micromobility solutions will continue enhancing the US transportation system over the next few decades. Public transport agencies and policy makers should continue collaborating with innovative micromobility companies, with the ultimate goal of promoting more sustainable and effective transport options.

KEY TAKEAWAYS
- Micromobility has emerged as a disruptive transport alternative offering more sustainable modes of transportation to private cars, while helping reduce traffic congestion in cities.
- Due to its relative affordability and convenience, micromobility options—including bikesharing systems and electric scooters—have increasingly been perceived as an effective first- and last-mile transportation method.

---

137 Kersten Heineke et al., "Micromobility's 15,000-Mile Checkup," McKinsey & Company, January 29, 2019.

- Public-private partnerships are being established toward the deployment of micromobility options. Cities are further promoting the implementation of infrastructure (e.g., bike lanes) necessary for a safe micromobility launching.
- Ensuring a uniform deployment of micromobility rideables remains a challenge for service providers. Nevertheless, cities and jurisdictions are increasing efforts to making bikes and e-scooters available and accessible to all.
- Micromobility can be an effective transport method in "transit deserts" and help bridge the gap between the level of transit services and the transport needs of communities.

CHAPTER 7:

MICROTRANSIT

———

"American cities are growing, traffic is getting worse, emissions are surging, and public transit systems are suffering from years of underfunding and neglect. On-demand, shared microtransit might gift entire cities with faster, cost effective service and move drivers out of personal cars."[138]

—*AARIAN MARSHALL,*

TRANSPORTATION REPORTER AT WIRED

According to data from the American Public Transportation Association (APTA), public transportation is ten times safer per mile than traveling by automobile.[139] Using public transit instead of a car reduces a person's chance of being in an accident by more than 90 percent. Moreover, the organization suggests a household can save nearly $10,000 annually by taking public transportation and living with one less car. Further, APTA claims communities that invest in public

---

138 Aarian Marshall, "LA Looks to Rideshare to Build the Future of Public Transit," *Wired*, October 24, 2017.

139 "Public Transportation Facts," American Public Transportation Association (APTA), accessed October 1, 2020.

transit lower the country's carbon emissions by thirty-seven million metric tons annually.

That being said, you're probably thinking "Wow—superior safety, affordability, and sustainability to cars," US public transit ridership levels must be skyrocketing! Wrong. Not even close. Despite a slight increase in the number of rides by half of a percent from 2018 to 2019, the truth is public transit is generally not perceived as an attractive alternative to the personal vehicle.[140] This explains, at least partially, the pressing traffic congestion issue witnessed all across the country's highways and roads. According to mobility analytics and connected car services company INRIX, Americans lose an average of ninety-seven hours a year in traffic, costing them nearly $87 billion or an average of $1,348 per driver.[141] Additionally, 45 percent of the American population still does not have access to public transportation.[142]

---

140 The half of a percent increase in the number of passenger rides from 2018 to 2019 corresponds to the computed average of the percent change for all four quarters from 2018 to 2019 (-1.70%, 0.46%, 2.20% and 0.90%) obtained from the following sources:
   Transit Ridership Report: First Quarter 2019, American Public Transportation Association (APTA) (Washington, DC: APTA, 2019).
   Transit Ridership Report: Second Quarter 2019, American Public Transportation Association (APTA), (Washington, DC: APTA, 2019).
   Transit Ridership Report: Third Quarter 2019, American Public Transportation Association (APTA), (Washington, DC: APTA, 2019).
   Transit Ridership Report: Fourth Quarter 2019, American Public Transportation Association (APTA), (Washington, DC: APTA, 2020).
141 "INRIX: Congestion Costs Each American 97 hours, $1,348 A Year," INRIX. press release, February 11, 2019, on the INRIX website, accessed October 1, 2020.
142 "Public Transportation Facts," American Public Transportation Association (APTA), accessed October 1, 2020.

Fortunately, innovative transport options powered by digital technologies are emerging across the nation and increasingly being perceived as valuable additions to existing mobility solutions. The hope is to make public transportation a more attractive and available option and thus increase its ridership levels. Microtransit is one of these transportation additions, offering first- and last-mile services, and feeding the main transit network. Microtransit is providing improved coverage in certain areas, in addition to more direct trips where the fixed routes require transfers.[143]

SAE International, previously known as the Society of Automotive Engineers, defines microtransit as:

"A privately or publicly operated, technology-enabled transit service that typically uses multi-passenger/pooled shuttles or vans to provide on-demand or fixed-schedule services with either dynamic or fixed routing."[144]

In the definition above, the term of "technology-enabled" is the real differentiator between microtransit and other shuttle services that have already existed for decades. A number of transport services providing traditional public transportation and demand-responsive systems have been developed by local authorities since at least the 1960s.[145] In a 2004 report,

143 Yann Leriche, "Microtransit: The Next Mobility Revolution or Much Ado About Nothing?," *Medium,* October 12, 2019.
144 "What is Shared Mobility?," Shared Mobility, SAE International, accessed October 1, 2020.
145 In the 60s and 70s, on-demand transport systems relied on dispatching centers receiving telephone calls from customers and assigning vehicles to service them:

the Transportation Research Board (TRB) identified more than fifty such transit systems in North America.[146]

MICROTRANSIT: A FIRST AND LAST-MILE MOBILITY SOLUTION FOR SUBURBAN AND RURAL AMERICA

Reflecting on the concept of microtransit, I recall relying on shuttle services to and from Washington, DC on more than one occasion during my first years upon arrival to the US. When I first came to the US, I was living at my aunt's house in Vienna (Virginia) from 2007 to 2009. During this time, I remember the Snowpocalypse winter storm that crushed the DC Metropolitan area in December of 2009. Over sixteen inches of snow created whiteout conditions at all surrounding airports. This event ultimately cancelled the first leg of my scheduled flight to Ecuador from Baltimore/Washington International Airport (BWI), leaving me stuck for six consecutive days.

Considering my aunt's fear of driving one hundred miles roundtrip at 4:00 a.m. on icy roads (I couldn't blame her), I decided for the first time to give the airport-shared van service SuperShuttle a try. The DC metro service was not an option that early in the morning, the Ubers and the Lyfts were still not around, and a taxi ride would have cost me at least $200. My journey was fairly simple and affordable. For $40, I reserved a seat in advance, the van picked me up

B. Arrillaga et al., *Demand-Responsive Transportation System Planning Guidelines*, Special Report 136 (Washington, DC: National Academy of Sciences - National Research Council, 1974).

146 David Koffman, *Operational Experiences with Flexible Transit Services* (Washington, DC: Transportation Research Board, 2004).

several hours prior to my flight's scheduled departure, and I was able to ride to the airport with three to four other people. While multiple stops to pick up other passengers certainly made the drive longer than a direct ride, it was worth it a million times over. It was a convenient and relatively cheap (saving me ~$160) innovative transport option for a long and hazardous ride to the airport.

Similarly, if you traveled to Las Vegas prior to 2015—before Uber and Lyft were allowed at McCarran International Airport—chances are you found $10 shuttle rides to be the most affordable transportation option to your hotel on the Strip. I personally made it to Vegas for the first time in 2013. I was immediately surprised to see how all arriving passengers would walk straight to the shuttle counter to purchase shared van tickets. Upon realizing almost all Vegas hotels were clustered together on the Strip, it all made sense to me. Why pay a taxi when you can, basically, split the cost of the ride with people going in the same direction and end up only feet away from your destination? Yes, I will admit my friend Sebastian and I were somewhat anxious to arrive to our hotel and start trying our luck at the casinos. Nevertheless, under a different context to cruising the 2009 Snowpocalypse on SuperShuttle, the convenience and affordability made it all worth it.

From my travel experiences above, microtransit for me has been defined as a technology-enabled system of multi-passenger shuttles. Considering the increase in smartphones adoption and the associated explosion of data coming from such devices, most microtransit solutions today will involve a mobile app or digital platform.

Coming from a communications background, Josh Powers walked me through his journey to become contract administrator and regional transit liaison between Kansas Johnsons county government and the Kansas City Area Transit Authority (KCATA). During our interview, Powers shared a few of the microtransit projects going on in suburban and rural areas of Kansas he has been involved with.

"I was coming out of a career in communications and public relations, and so I ended up working for the Kansas Department of Transportation in 2010 as their person on public relations and representing the Secretary of Transportation."

"When they were developing transportation programs through the state of Kansas, staff, including myself as a communications person, would go out to all one hundred five counties in Kansas and talk. We would do this week-long tour of the state to meet with stakeholders and hear what they wanted to see in the transportation program."

During his six years at KDOT, Powers saw a complete rebuild of the statewide public transportation organization. As such, the Coordinated Transportation Districts (CTD) was created to make sure all of the service providers in the Kansas transit program were talking to each other rather than offering duplicative services.

"So you had small towns with only one dialysis center within five hundred miles, and all these different providers making their own trips with one person on a vehicle. We can get these folks to work together, you can get ten people on a vehicle

and be much more efficient. So, we did that, and it was very successful and saved a lot of money."

After looking at a position as the State Public Transportation manager, Powers saw the opportunity to take over the system for Johnson County (Kansas) in 2016. The KCATA formed a partnership with Johnson County, Wyandotte County, Kansas City (Kansas), and the City of Independence (Missouri), among other jurisdictions, to create one single regional transit system called RideKC. Contrary to the previous very disjointed system where individuals had to constantly switch bus lines and brands, travel in Kansas was now streamlined.

"And so now what we had is a one branding for seven different systems, and it's now regionally coordinated. So, it's a seamless network now."

Powers underscores the need for a robust public transportation system in Kansas, especially due to its low population density. He claims transit service levels are, in fact, very low and people may need to walk more than a half mile to get to a bus stop, and then deal with thirty-minute headways in the best-case scenario. For people facing a disability, it can be really hard to even make it to that first stop.

In light of this, he says the benefits new microtransit initiatives have brought to the first- and last-mile challenge in the city are just the beginning of a solution. Powers recognizes there is a challenge of serving such a large area with a small fleet of shuttles. He speaks to a partnership created with a taxi company to mitigate more potential capacity problems.

"We only have a fleet of seven vehicles, and we are sixty square miles, six hundred thousand people, so you run into capacity issues. So, what we've done is we have a contract with a local taxi company."

All deployed vans comply with the Americans with Disabilities Act (ADA). As part of this innovative microtransit effort, they have partnered with software company TransLoc Inc., whose mobile app has to be downloaded to request a shuttle. Powers explains the efficiency and affordability of RideKC while speaking to the disconnect in ridesharing services like Uber with serving the rural population travel needs.

"Those drivers for Uber want to prioritize those short trips that are in the urban core so they can get a person and get them off, and get the next person to make money... You can dial up an Uber in Johnson County, but that's going to need twenty minutes of deadhead time for the driver to get there. So, he's already disincentivized from coming because one trip is going to take up a lot of his time. And then, when he gets that person and if that person is going downtown, then it's twenty minutes back and it's going to cost the passenger $25 to $30 for that trip."

"Our microtransit will get you close to that same trip and connect you with a fixed route that gets you where your final destination is for $1.50. So, it's a pilot, and we expect that fare of it will not always be $1.50. It's $1.50 because that's what our fixed route fare is, and we want to be standardized across the system."

The reason for such a low price is, in fact, to incentivize people to use microtransit and to connect to fixed route transit.

Powers says RideKC is increasingly being perceived by the people of Johnson County as an efficient mobility solution.

This should not be surprising. Just like we spoke about micro-mobility services as a potential solution for first- and last-mile trips in transit deserts, a microtransit solution can more effectively bridge that transport gap between residential or job locations and fixed route transportation. More so, the wider the array of transport offerings we can make available to the population, the better.

Speaking again to the importance of having access to a robust public transportation system, Josh Powers highlights the advantages of microtransit in a *JoCo on the Go* podcast episode.[147] He explains how a pilot program is helping Johnson County residents get from point A to point B in a new, convenient, and affordable way.

"It [public transportation] really does touch every segment of our community. People use public transportation to get to work, they use it to get to medical services, they use it to go shopping, and just participate in their community. Here in Johnson County, what that really looks like, for the most part, is commuter service, so folks leaving the county and going to the urban core to work and then coming back again in the mornings and the evenings."

"So we've known for a long time intra-county service is limited; people want to be able to get east and west. If you're in Lenexa

147 Josh Powers, "Microtransit," September 8, 2019, in *JoCo on the Go: Everything Johnson County Kansas,* produced by Theresa Freed, podcast, Podbean, 15:39.

it can be difficult to get [five miles away] to Overland Park, and so forth. So that's really what was part of the impetus for microtransit, to be a new service that connected people to fixed route service but also gave them better mobility options."

Powers also refers to the level of support they've received from the Board County of Commissioners (BOCC), who have pushed their staff to be innovative and tackle the issue of access to public transportation. Microtransit is the "sweet spot" between ride-hailing ridesharing services like Uber and Lyft, and the first- and last-mile connection solution.

"You can use it in either way, that is to get to a bus route to go on, or you can go directly to the front door of what your destination is."

The travel booking process is also similar to Uber and Lyft, where customers use their phones to access the service which, in Powers words, is "people's instantaneous gratification". As such, users create an account, enter their payment details, select the destination, and have the app provide them an estimated time of arrival of the driver and the duration of their trip.

Powers touches on RideKC's established partnerships, like the one with Johnson County Community College (JCCC) who have enrolled in their U Pass program. The school provides subsidies so students, faculty, and staff can ride for free with a valid JCCC ID. Another partnership they have is with a farmers' market in downtown Overland Park, where the market pays for any trip on Saturday that goes anywhere downtown. The microtransit service is perceived as a convenient solution to an apparent parking problem around the market.

"So people like that front door access, and microtransit gives that to them without having to worry about where they're going to park or how long, how far they can walk with their purchases."

Powers underscores that the idea is to provide a convenient and affordable mobility option for the people of Johnson County.

Following Powers interview, Johnson County BOCC vice chair and supporter of microtransit, Jim Allen explains how a next step will consist of identifying the routes where usage is low and adjusting them as a way to increase revenue. This would certainly require big data and analytics solutions.

Allen suggests the microtransit pilot program has had great success at helping the fixed routes, especially increasing transit ridership across the county. The host shares a testimonial of a Johnson County microtransit driver, who underscores the economic and societal benefits of such service for people of all ages with different needs.

"We get a lot of students from JCCC and the other schools in the area, and also some of our high schools. We also help out with people trying to get to work in the area. People are also trying to make medical appointments for elderly people, so we do a lot with the elderly people and we also are capable of taking a wheelchair for the wheelchair dependent riders."

"People are ecstatic about the service; they feel like it's one of the best things that could ever happen to them. It's a nominal charge of $1.50, so for people who are economically strapped and can't afford to live in the Uber this has really come as a godsend for them when they rave about that."

As I was listening to the testimonial above on transport advantages for the elderly people and their medical appointments, I remembered I had come across a similar mobility service. It was called GO!Bus Plus, a six-month microtransit pilot program launched in July 2020 in Grand Rapids, Michigan. This program allowed riders to reserve or call for rides when needed through a mobile app and web portal with various transportation providers. The program aimed to streamline the process of scheduling medical transportation and was a partnership between distinct stakeholders. Such partnership ecosystem was comprised of public bus transit system The Rapid, healthcare logistics platform Kaizen Health, disability advocates of Kent County, and the city of Grand Rapids.[148,149]

Given the convenience and affordability offered by microtransit solutions in US suburban and rural communities, it is no surprise similar transport services are being implemented in other regions of the world.

---

148 "Pilot Program to Boost Mobility Options for People with Disabilities," Mibiz, July 13, 2020.

149 "New App Offers The Rapids GO!Bus Passengers Convenience and Less Wait Time," The Rapid. press release, August 5, 2019, on The Rapid website, accessed September 8, 2020.

Yann Leriche, former CEO of French-based mobility operator Transdev, discusses how Chronopro—a microtransit solution launched by Transdev and its subsidiary Cityway—in 2016 entered the French southern city Vitrolles with a population of thirty-four thousand.[150] Leriche suggests the on-demand shuttles provide a reliable connection from a transit hub to office buildings within a nearby office park. He further explains how Chronopro vehicles have come to replace two local fixed bus routes, previously used to serve the locations within the business cluster.

"The microtransit vehicles are well coordinated with the real-time arrival of buses at the transit hub guaranteeing passengers a seamless connection every time. Shuttles to and from the transit hub can be booked through an app, a website page, or via a call center. Routes are automatically computed according to demand to bring each passenger right to their destination within the office park service area."

Leriche affirms the Vitrolles microtransit solution has been very successful compared to preceding fixed route service. In fact, it has reduced average travel time by 48 percent, increased ridership by 42 percent, lowered miles traveled by 84 percent due to a more efficient route service, and expanded the number of destinations/stops served by 80 percent.[151]

---

150 Yann Leriche, "Microtransit: The Next Mobility Revolution or Much Ado About Nothing?," *Medium*, October 12, 2019.
151 Ibid.

REVIVING URBAN PUBLIC TRANSIT

Via is a mobility company that develops innovative solutions for on-demand and prescheduled transit, powered by technology. With a more urban focus, it is another relevant example of the deployment of on-demand services. In 2019, the Central Ohio Transit Authority (COTA) partnered with Via and launched the COTA Plus microtransit network.[152] Likewise, COTA launched a second on-demand venture with Via in October 2020 to provide the first full-time on-demand bus service in the US.[153] Bus? Where is the "micro" in a bus? What happened to the shuttles and vans? Keep calm and call it microtransit for mass usage.

In the first three weeks of service, COTA/Plus Bus On-Demand (the new transit system's name), customers had an average wait time of thirteen minutes to be transported to any transit stop within their defined zone. Daniel Ramot, Via CEO and cofounder explained:

"Dynamic, on-demand transit is a powerful tool to expand access to professional, economic and social opportunities for communities in the Columbus region... We are proud to work with COTA to introduce an innovative transit solution, powered by technology, that complements and extends the existing public transit infrastructure, and meet the needs of residents with increased efficiency, affordability, and convenience."[154]

---

152 "COTA, Via Partner to Provide On-Demand Service in Central Ohio Communities," *Mass Transit*, October 15, 2020.

153 Ibid.

154 Ibid.

Two years before, Ramot discussed the future of transportation at the 2018 Web Summit, an online technology event bringing together the people and companies redefining the global tech industry.[155] He shared his vision for where we are headed and how shared mobility through vans and shuttles can be integrated into cities legacy public transit system.

Ramot speaks to the poor economies of scales of the personal car, claiming 85 percent of people commute by car, and of those 90 percent commute alone. Consequently, road infrastructure becomes easily overwhelmed. Via's mission is to solve this problem of too much demand for a very limited infrastructure.

The deployments of buses have, in fact, not effectively addressed this issue over decades; one of the causes being buses run on fixed routes and schedules. Interestingly enough, Ramot makes reference to a study by the Brookings Institute—a Washington, DC-based research group—suggesting as the quality of the public transit system improves, people utilize public transportation more and fewer people own a car.

Ramot sees the solution to traffic congestion in dynamic on-demand shuttles that can effectively serve people much better than traditional buses. He thinks regular transit buses are only efficient to commuters if they live or work within a certain walking distance, a certain radius from one of the

155 *Web Summit*, "Moving Past Private Vehicles and Public Buses," November 6, 2018, video, 11:32.

bus stops. As we know from Josh Powers, this can require people to walk large distances often.

"With a dynamic shuttle, what you could do is replace that [bus] system or complement it if the bus is sufficiently efficient already with a very dense network of what we call dynamic virtual bus stops that the system can direct you to walk to."

"You can also book a van that can take you from any virtual bus stop to any other virtual bus stop, and you're always within, perhaps, a few hundred meters at most of any one of these virtual bus stops. The system is much more agile, relies on the smaller vehicles, is asset light, and fundamentally doesn't require any new infrastructure. In fact, the infrastructure, we would like to say, is the data that allows the system to become increasingly more efficient."

Over the last few years cities are increasingly becoming interested in these types of solutions. Before, Ramot says, he would meet with the city and tell them about the company's vision and how to leverage software already working well in New York. Yet, they would just thank him for coming and that is how far they would get.

The situation is different today, and many cities including Singapore, London, and Sydney are actively experimenting microtransit, dynamic, on-demand transit solutions. This is because the infrastructure is overwhelmed and congested. According to Ramot, building a subway solution is one alternative, but an extremely expensive one and not for short-term purposes.

Speaking to a few technological challenges for Via, Ramot suggests running a system with computations for dynamic virtual stops is much more complex than for a commute journey from point A to point B. There is an associated challenge with ensuring optimum seat utilization, critical for maintaining the cost per seat down and thus justify the solution as a public transit system. When a user requests a shuttle, the system could assign them tens of vans that have multiple passengers and assign thousands of different route solutions. Ramot claims Via has seen better operational metrics than traditional bus systems. This includes key metrics like demand utilization, wait time, and cost per ride.

Lastly, Ramot refers to the key role Via's partners have played into deploying the company's microtransit solutions. More specifically, he discusses how their partnership with Daimler, and especially Berlin's local public transit authority BVG, allowed Via to launch a van service in the German capital. The service required a special permit since the operation of shared dynamic transit services like Via was not allowed by Berlin regulations. As such, the role of policy makers was key for the successful implementation of an innovative microtransit service for Berlin's residents.

"The BVG was able to convince the Berlin Senate to provide us with a special license to operate the service, and about a couple of months ago we launched this service called BerlKönig. It is the largest publicly operating dynamic transit fleet that's connected to a local public transit service. It is also the largest such electric fleet, 80 percent of this fleet in Berlin is electric."

"And I think it's a beautiful example of how a company that's bringing a new technology to the market can work with the city and the transit authority to launch an innovative service. It can complement the existing services, the bus, the tram, the subway, and all of the services the BVG is operating extremely effectively in Berlin. We can add to that and provide a solution for when you need to get from [point] A to between two points that aren't well served by the existing public transit service."

Summarizing the value to customers of recently emerging microtransit services, it comes down to two key elements: providing an enhanced experience and improving efficiency. If you think about it, this is at least part of the reason that has enabled ride-hailing ridesharing companies like Uber and Lyft to attract new customers. Unfortunately, many long, well-established taxi services missed that digital train.

Consumer-centric transport options reminds me of my transportation project in Mumbai (India). My team and I explored this emerging trend of tech-driven bus and shuttle services penetrating India. Mid- and long-distance bus service company Shuttl ran over two thousand buses in six Indian cities, and allowed riders to book their trip through a mobile app showing routes, stops, and seats.[156] These shuttles offered superior convenience like air conditioning, as opposed to the old-fashioned buses from the rather obsolete Indian public transit.

---

156 "Shuttl: Stress-Free Commute to Work," Shuttl, accessed October 5, 2020.

KEY TAKEAWAYS

- Leveraging data and technology, microtransit appears as a cost-effective, sustainable, convenient, and inclusive transportation method for rural, suburban, and urban US communities.

- In suburban and rural areas mainly, microtransit vehicles are seen as effective additions to existing underfunded transit systems, serving as first- and last-mile connections and enabling people access to jobs, resources, and other social privileges.

- Predominantly in urban areas, microtransit is helping reduce personal car driven miles, and associated traffic congestion and vehicular carbon emissions.

- Microtransit initiatives are coming from public-private partnerships, involving the collaboration of distinct stakeholders including policy makers, transit agencies, and mobility and technology companies.

CHAPTER 8:

SHARED MOBILITY AND MOBILITY AS A SERVICE

"Desperation sometimes drives innovation."[157]

—DARA KHOSROWSHAHI, UBER CEO

According to United Nation's (UN) data, the world's population is on pace to add another 2.5 billion people to urban areas by 2050. In the case of the United States only, the urban population would see an increase of approximately seventy million people, augmenting the proportion of urban residents from 82 percent (in 2018) to 90 percent.[158]

Earlier in the book, I referred to a myriad of problems associated with the current US transportation system. More so, our urban future sees the already hectic traffic congestion, excessively long commute times, involved vehicular crashes,

---

157 Leslie Hook, "Expedia Boss Dara Khosrowshahi on the New Breed of Disrupters," *Financial Times*, July 23, 2017.
158 "2018 Revision of World Urbanization Prospects," Department of Economic and Social Affairs, United Nations, May 16, 2018.

and vehicular air pollution compounding as the population grows.

Bringing more people to cities comes with an increase in demand for transportation, and present Americans' dependency on personally owned cars is quite evident. As mentioned previously, the United States has the highest rate of per capita vehicle ownership among major countries with 838 per 1,000 Americans.[159] Likewise, vehicular carbon emissions are already a major concern; which a rise in cars on the streets would only aggravate and delay the arrival of a definitive solution to the climate change problem.

Fortunately, as ride-hailing company Uber CEO Dara Khosrowshahi has interestingly manifested, desperation can lead to innovation. Desperation to prevent vehicular traffic from completely collapsing our cities. Desperation to save the environment for us, our loved ones, and for future generations. Desperation for an alternative to expensive taxis and old-fashioned buses. Indeed, desperation is, at least partially, what has triggered our imagination and creativity in light of the transportation sectors improvement.

As such, shared mobility and Mobility as a Service (MaaS) are emerging trends serving as innovative mobility solutions to the rapid US urbanism and associated increased traffic congestion and environmental pollution.

---

159 "State Motor-Vehicle Registrations 2018," United States Department of Transportation Federal Highway Administration, Policy and Governmental Affairs Office of Highway Policy Information, December, 2019.

SHARED MOBILITY

Let us start with the concept of shared mobility, which has gained impressive momentum over the past few years. SAE International defines it as:

"The shared use of a vehicle, motorcycle, scooter, bicycle, or other travel mode. Shared mobility provides users with short-term access to one of these modes of travel as they are needed."[160]

Further, the technological advancements have made shared mobility a demand-driven transport alternative. Travelers can share a vehicle either over time (e.g., bikesharing, e-scooter) or simultaneously. In earlier chapters we touched on shared transportation services like micromobility and microtransit. We will now look at the broader category that also includes traditional ride-hailing and carsharing services.

According to a research report by Facts & Factors Marketing—a leading market research organization—the global shared mobility market will reach $238.03 billion by 2026 from an estimated $99.08 billion in 2019, expected to grow at a compound annual growth rate (CAGR) of 15.42 percent.[161]

I have to admit Uber and its competitor Lyft have made my commuting journeys much easier, and chances are they have impacted yours, too. The concept of simply pulling up your phone and requesting a ride through a mobile app is

---

160 "What is Shared Mobility?," Shared Mobility, SAE International, accessed October 1, 2020.

161 "Global Shared Mobility Market Size & Trends Will Reach to USD 238.03 billion by 2026: Facts & Factors," Facts & Factors, GlobeNewswire, December 10, 2020.

convenient at the very least. An app that displays the estimated time of arrival for the driver heading to your pickup location, and that notifies you when the driver is about to arrive is quite an innovation and models convenience, if you ask me. On top of that, having the fare automatically calculated and charged to the credit card you've linked to your Uber or Lyft account has made my trips seamless. Likewise, do I need to mention the sometimes underestimated benefit of knowing the name of the driver and the car's plate number and rating before of jumping into the vehicle?

Additionally, consider the advantage of saving time, money, and unnecessary stress when looking for a parking spot in the city. Moreover, I also think of the positive impact growing adoption could have on reducing impaired driving. Not too long ago, before the Ubers and the Lyfts showed up, we didn't really have much of an efficient door-to-door transport alternative—other than traditional taxis—to driving our personal cars for a night of drinking with family and friends.

Nonetheless, and as you may be correctly thinking, ride-hailing itself does not directly tackle the major transport challenges of long commutes and contamination of the environment. Interestingly enough, cofounder and former Uber CEO Travis Kalanick describes during a TEDx Talk in 2016 how Uber didn't start out with grand ambitions to cut traffic congestion and pollution. Yet, as the company was launched and evolved, he wondered if there was a way to get people using Uber along the same routes to share rides, reducing costs, and their carbon footprint along the way. The result: UberPool, the company's carpooling service, which in its first eight months took 7.9 million miles off the roads and 1,400 metric tons of carbon dioxide out of the air in Los Angeles.[162]

"The beginning of Uber in 2010 was we just wanted to push a button and get a ride. We didn't have any grand ambitions. But it just turned out that lots of people wanted to push a button and get a ride, and ultimately what we started to see was a lot of duplicate rides. We saw a lot of people pushing the same button at the same time going, essentially, to the same place."

"And so we started thinking about how to make those two trips into one. Because if we did, that ride would be a lot cheaper, up to 50 percent cheaper. And, of course, for the city you got a lot more people and fewer cars. And so, the big question for us was would it work? Could you have a cheaper ride, cheap enough that people would be willing to share it? And the answer, fortunately, is a resounding yes."

Kalanick suggests the effect of UberPool in San Francisco was an increase in the number of people getting around the city and a reduction in the number of cars on the road. This positive outcome led the company to implement the same carpooling service in Los Angeles where, aside of the success at reducing miles off the roads and decreasing carbon emissions, Uber successfully added one hundred thousand new people carpooling every week.

---

162 *TED*, "Ubers Plan to Get More People into Fewer Cars | Travis Kalanick," March 25, 2016, video, 19:18.

I have personally used UberPool on countless occasions and the benefits are very tangible. Having the fare substantially decreased does make a big difference and creates an important additional incentive to adopt the ride-hailing, carsharing company as a preferred method of transportation. Yes, you may be sharing your ride with complete strangers who, if passed 2:00 a.m. on a Friday night, may make a show of themselves. But remember, sharing is caring, and in this case, sharing is caring for the environment and for our cities' transportation planning and drivability. Indeed, even if my rides may be longer because the car has to make a few extra turns to pick up other passengers, the added time is worth it in most occasions.

Considering the financial and convenience advantages ride-hailing and carsharing services have brought to the American population, it should not surprise us their market has significantly increased over the past several years. In 2015, the ride-hailing market comprised twenty-five million riders in the United States and Canada, or roughly 7 percent of the addressable population.[163] This market increased in 2018 to sixty million riders, with an addressable population penetration rate of close to 18 percent.

As mentioned earlier, the US has one of the highest rates in the world of per capita vehicle ownership. According to US Census Bureau data presented in a Nasdaq article, the percentage of vehicle-less households in the US has steadily and significantly decreased since 1960, as the American Dream

---

163 Luke Lango, "The Era of Car Ownership Is Over. And These 4 Charts Prove It," InvestorPlace, April 3, 2019.

has constituted owning a car. From 1960 to 2010, the percentage of households without a car dropped in the US from about 22 percent to 8.9 percent.[164]

Yet, as a result of the aforementioned benefits offered by new mobility solutions, car ownership rates in the nation are now dropping for the first time in modern history. From 2010 to 2015, and for the first time since 1960, the percentage of vehicle-less households rather increased to 9.1 percent.[165] This five-year period coincides with the penetration of ride-hailing, carsharing services presented above.

This brings us to the question of whether the trend of car ownership in the United States is starting to decline. Other than ride-hailing and shared mobility services, what is influencing this interesting phenomenon of people giving up their vehicles?

MOBILITY AS A SERVICE (MAAS)

If you recognize and enjoy the advantages of ride-hailing and shared mobility, chances are you will find Mobility as a Service (MaaS) even more appealing. Of all the solutions we have talked about so far, I consider MaaS to be the most likely to solve the transportation system problems in the US. Simply put, it is the future of transportation.

Mobility as a Service (MaaS) is a relatively new concept first presented in 2014 during the European Congress on

---

164 Ibid.
165 Ibid.

Intelligent Transport Systems in Helsinki (Finland). MaaS incorporates the technological concepts we have discussed to this point (e.g., big data, AI, cloud) and the aforementioned revolution in on-demand transportation systems microtransit, micromobility, and ride-hailing. As such, MaaS consists of a movement from personally owned vehicles toward mobility solutions provided by a combination of public and private organizations.

Given the recent appearance of the concept of MaaS, there is no single official definition for it.

The UK's House of Commons Transport Committee defined MaaS in 2018 as:[166]

"Digital platforms through which people can access a range of public, shared and private transport, using a system that integrates the planning, booking, and paying for travel."

Indeed, by integrating distinct transport services (e.g., bus, metro/train, car, bike, scooter) into a single digital platform on the "single-ticket principle," MaaS offers personalized transport plans tailored to customer needs. The goal is to make it so convenient for users to get around they choose to give up their personal vehicles for city commuting, not because they're forced to, but because the alternative is more appealing.

---

166 *Mobility as a Service*, House of Commons of Transport Committee (London, U.K.: authority of the House of Commons, 2018).

As such, you might compare Mobility as a Service to Movies as a Service, or simply refer to it as the "Netflix of Transportation."[167] Before Netflix, chances are you would watch a movie—in DVD or VHS format—by renting it from Blockbuster, or by taking a trip to the closest AMC or Cinemark. Today, Netflix allows you to search for all available movies, select the genre(s) of your choice, and pay for it in a single subscription. You can watch anywhere you want, anytime you want, all in the same platform without having to jump from one app to another. Similarly, MaaS is about combining all available transportation modes into a single digital platform to provide the customer a seamless experience.

In November 2017, after nearly a year of testing, MaaS Global, a private startup based in Helsinki, launched its Mobility as a Service mobile application *Whim*. In late 2018, the Whim app had over seventy thousand registered users.[168] MaaS Global claims to be the world's first true MaaS operator, interconnecting many of the city's mobility options under one subscription. A single app allows users to combine, plan, and pay for public transport, taxi, car rental, carsharing, e-scooters, and city bike trips. Anyone with the app can enter a destination and select their preferred mode of transportation, or a combination can be used in cases where no single transport mode covers the door-to-door journey. Users have the option to pre-pay for the service as part of a monthly mobility subscription or pay as they go using a payment account linked to the service.

167 Andy Boenau, "Mobility-as-a-Service 101," *blog,* accessed October 5, 2020.
168 Ari Hartikainen et al., "Whimpact," *Ramboll,* May 21, 2019.

As of January 2021, five different service tiers exist: a pay-as-you-go and multiple prepaid plans ranging from $80 to $800 (unlimited) for a month of service.[169] Whim is available for commuters with various personal transportation needs. These needs are covered by plans such as pay-as-you-go with no monthly subscription fee, to monthly subscription packages with unlimited number of public transportation trips, car rentals, thirty-minute bike rides, and up to eighty discounted taxi rides. Whim is also present today in select cities in the UK (Birmingham), Austria (Vienna), Japan (Tokyo), and Belgium (Antwerp).[170]

I had the pleasure of interviewing Sami Pippuri, former chief technology officer (CTO) at Global MaaS and creator of the Whim app. At the start of our conversation, he shared with me his previous experience at mobile company Nokia and expressed his view on the similar technology disruption in the telecom and transportation sectors.

"The mobility side of things came a little late because I had been working with telecoms and mobile technology. But that's probably not a coincidence because I'm based in Finland and it's the home of Nokia, and in the turn of the millennium it was the leader in the global industry actually creating the mobile industry."

"And a lot of the same sort of mega trends that made that breakthrough in telecom possible, and which I was involved in, are actually hitting transportation as well."

---

169 "Find Your Plan," *Whim*, MaaS Global Oy, accessed January 15, 2021.
170 Ibid.

Pippuri spent twelve years working at Nokia in different kinds of roles, including software development and hands-on coding. He claims some of his software code is in hundreds of millions of devices out there. He also worked in cybersecurity and media and thinks MaaS Global's founders would not have reached out to him without his previous professional experience.

Interestingly enough, he confessed before joining MaaS Global he was incredulous about the whole concept of Mobility as a Service and the intended idea of a shift toward multimodal transportation and away from private car ownership.

"So one day I got a call from one of the founders of MaaS Global saying they would like to talk to me, and they have this concept called Mobility as a Service, and I had read about it in the news. Someone wants to give everyone all-you-need taxi rides for $100 a month, and I thought, 'I don't know, I kind of like cars.'"

"So I wasn't sure going into that meeting, but that night I actually got to talk to his partner at the time, to know what was involved and what it was they really wanted to achieve. I thought, 'this is going to happen anyway because that's just inevitable. So, if it's not these guys, somebody else will do it, and might as well be them.' So, I got on board."

Moreover, Pippuri described the humble beginnings of MaaS Global. The company started in a very small office with some funding to create a Minimum Viable Product (MVP), which they would launch only three months later. This would allow MaaS Global to move to the next level and start working more on the product and distinct value propositions. Their first assumption, Pippuri said, was people would want a set number of credits they could spend on different modes of mobility. Yet, it turned out to be a bad idea as people didn't like it. Nonetheless, Pippuri suggested the startup was able to pivot from that quickly and have a more understandable value proposition where taxi rides and City bike rides were included. Like that, people did not have to think about balances and credits they had consumed.

Further, Pippuri made an analogy with the telecom industry and how, back in the day, it used to be a pay-as-you-go type of pricing for, say, two hundred to three hundred minutes a month. Today, on the other hand, every mobile operator offers unlimited packages and thus users don't need to worry about how much they speak in a month.

"So that was kind of the idea there, as well, that you have to get to a place where the packages are enough, and people just don't worry about it anymore, and there is no sticker shock at the end of the month. 'Oh, wow, who in my household was riding this much taxi' or something like that."

In addition, Pippuri referred to the elevated cost of car ownership and a "disconnect" when someone spends a high amount of money (e.g., $50,000) every four to five years on an asset that is parked most of the time. It thus becomes wasted money, he claims. As such, and with the emergence of ride-hailing and ridesharing services, Pippuri explains how in large US cities people are increasingly accepting giving up on their vehicles.

"In some cities like Los Angeles, for example, there are a lot of people who don't own a car, and ten to twenty years ago that would have been unheard of, but now it's kind of a cool thing. You have UberPool, or you just have Uber and that says you don't need a car."

"So, I think Mobility as a Service probably takes a slightly different form in the US but there are a lot of common elements and it's really centered around the ownership. Do you actually need to own a car? You can still drive one whenever you need it, but you don't have to park it in your garage."

MaaS Global's Whim mobility app is arguably one of, if not the most representative example of a Mobility as a Service solution. The company is widely recognized as a first mover in the MaaS industry and has been awarded with early brand recognition and substantial funding rounds, with nearly $36 million in late 2019 from global players such as BP Ventures and Mitsubishi Corporation.[171] Besides MaaS Global, however, there are other MaaS pilots and services—both private and public-driven—putting efforts toward the implementation of MaaS in Europe.

This is the case of *Jelbi*, a mobility app launched in September 2019 and run by Germany's fast-growing capital, Berlin's public transport authority Berliner Verkehrsbetriebe (BVG), and by Lithuanian mobility startup Trafi's technology. As a side note, Trafi has built a similar mobility platform in Vilnius (Lithuania) and has partnered with Prague (Czech

---

171 "MaaS Global Completes €29.5M Funding Round," *Whim (blog)*, MaaS Global Oy, November 7, 2019.

Republic) and Jakarta (Indonesia).[172] In an interview, Jelbi head Michael Heider described the mobility app as:

"A solution for people to move to shared forms of mobility—and to avoid a transport collapse."

Comparable with Whim, the Jelbi app integrates different transport modes together and allows its users to book and pay for their trips in a single platform. Means of transport and Jelbi's partners include tram and bus services, rail operator Deutsche Bahn, electric kick scooters from Tier and Voi, bikesharing from Nextbike, carsharing service from Miles, Flinkster, and Greenwheels, and on-demand minibus ridesharing service from BerlKönig.[173] Heider encourages the use of his company's MaaS solution:

"People should leave their cars behind and use Jelbi to get around town… We want Jelbi to be the number one interface for shared mobility in Berlin."

By September 2019, Trafi had raised a significant fourteen million dollars of investor support for Berlin's Jelbi deployment, certainly due to the mobility app success in Vilnius since its launching in 2017 with one-fifth of the population using it.[174]

---

172 Douglas Busvine, "From U-Bahn to E-Scooters: Berlin Mobility App Has it All," *Reuters*, September 24, 2019.

173 "The Jelbi Mobility Partners," Jelbi, accessed January 5, 2021.

174 Douglas Busvine, "From U-Bahn to E-Scooters: Berlin Mobility App Has it All," *Reuters*, September 24, 2019.

In February 2021, Trafi announced it will be taking Mobility as a Service to Latin America, starting with Bogotá, Colombia.[175] The platform will pull together public transit, including buses and trams, in addition to local taxis and electric bikes.

Similar to Trafi, there are other companies in the race toward the deployment of MaaS solutions. German Hamburg-based startup Wunder Mobility, for instance, has customers including automotive giants BMW, Volkswagen, Daimler, and Toyota, and has raised sixty million dollars in backing from investors to expand its platform in the US.[176] Managing partner David Blumberg from venture capital firm Blumberg Capital has expressed:

"The future of transportation is increasingly driven by smart, flexible, inter-operable mobility services."

Other examples of MaaS solutions include Upstream in Vienna and Graz (Austria), Transdev in Saint-Etienne (France), Ubigo in Stockholm (Sweden), and Moovit across cities in Italy, Portugal, and Spain.[177]

Moreover, although Mobility as a Service appears to focus mainly on urban areas, MaaS services can also be developed for suburban and rural residents as well as for other

175 Kirsten Korosec, "Trafi Takes its Mobility-as-a-Service Platform to LatAm, Starting with Bogota," TechCrunch, February 15, 2021.
176 Paul Sawers, "Wunder Mobility Closes $60 Million Round to Expand its Urban Transport Platform in the US," Venture Beat, September 19, 2019.
177 Mehdi Essaidi et al., "The Future of Mobility as a Service (MaaS): Which model of MaaS Will Win Through?," Capgemini Invent, Capgemini and Autonomy, 2020.

applications.[178] For instance, Yumuv is a regional MaaS mobile app powered by Trafi, integrating transport modes from Swiss cities Zurich, Basel, and Bern. It allows people to utilize, under a single subscription, public transportation and different micromobility brands (e.g., Tier, Voi) in all three cities, thus providing a "holistic multimodal service."[179] Further, the rural MaaS KomILand pilot projects allow users in three small Swedish villages to plan and book trips in an app, which offers a menu of transport options including ridesharing, public transportation, carsharing, and cargo bike rentals.[180]

MAAS PILOT PROGRAMS IN THE US

While Mobility as a Service has primarily emerged in Europe—as described through the numerous above MaaS solutions—I was pleased to learn about a few smaller scale MaaS pilot programs occurring recently in the United States.

During an *ITE Talks Transportation* podcast episode, Vincent Valdes, associate administrator for research, demonstration, and innovation for the Federal Transit Administration

---

178 Jana Sochor, *Piecing Together the Puzzle Mobility as a Service from the User and Service Design Perspectives,* International Transport Forum Discussion Papers, No. 2021/08 (Paris, France: OECD Publishing, 2021).

179 Martynas Gudonavičius, "Yumuv—the Next Big Leap for Mobility as a Service," Trafi, August 25, 2020.

180 Jana Sochor, *Piecing Together the Puzzle Mobility as a Service from the User and Service Design Perspectives,* International Transport Forum Discussion Papers, No. 2021/08 (Paris, France: OECD Publishing, 2021).

(FTA) from 2008 until 2020, shares how on-demand integrated mobility has been gaining momentum in the US.[181]

Valdes suggests companies across distinct modes of transport are collaborating among them and escaping from their respective niche sectors isolation and moving toward a Mobility as a Service ecosystem.

"It's an area that's evolving on a daily basis, and certainly one that we've had our finger on the pulse of for years. What is particularly important is the evolution in mobility. This is particularly exciting because it's a real paradigm change for the industry."

"We're seeing the traditional silos between all the different transportation modes have somehow become a little blurrier over the years. They're blurrier than they've been before, and that's a good thing for travelers everywhere."

Valdes discusses his involvement with MaaS since 2016 and refers to the launching of the Mobility on Demand (MOD) Sandbox Project from 2017 through 2019.[182] The purpose of this initiative, Valdes explains, was to truly investigate the idea of integrated transportation systems and observe how other emerging modes of transport could be coordinated together. As such, Valdes and his team encouraged transit agencies to better synchronize their services with other

181 Vincent Valdes, "The Impact of Innovation on MaaS/MOD with Vincent Valdes," *ITE Talks Transportations Tracks,* produced by Bernie Wagenblast, podcast, Spreaker, 19:05.

182 Mobility on Demand (MOD) and Mobility as a Service (MaaS) tend to be used interchangeably in the US.

transportation modes—such as microtransit or micromobility—by partnering with scooters or bikeshare companies already present in numerous US cities.

Furthermore, Valdes mentions his team has recently developed an Integrated Mobility Innovation (IMI) program which supports projects that analyze how new technologies like artificial intelligence (AI) can be leveraged to improve coordination between transit agencies and private sector mobility companies. The ultimate goal, Valdes says, is to provide a seamless transportation experience to all individuals.

"I think travelers have become very sophisticated and are not interested in mode specific transportation per se. What they're most interested in is thinking about that complete trip...how do I get from point A to point B in the most efficient, cost effective, comfortable way that I can, given my circumstances in the moment? It's a reflection, I think, of our societal preferences and certainly one we wanted to help transit accommodate."

Valdes claims the IMI program was indeed supporting twenty-five different projects that were looking at these relationships between distinct transportation modes.

In January of 2021, the Regional Transportation District (RTD) of Colorado announced the launch of a mobile app called FlexRide that fully integrates regular bus and rail services. Developed in partnership by the Finnish Mobility as a Service company Kyyti Group, FlexRide is an on-demand microtransit service for first- and last-mile connections.[183]

THE FUTURE IS INTEGRATED

In short, the concept of Mobility as a Service has been gaining more popularity over the last few years, with an increasing number of companies and startups penetrating this space. As such, according to a Research and Markets report, the global Mobility as a Service (MaaS) market is forecasted to reach the valuation of $280.77 billion in 2027 from $52.3 billion in 2019, at a CAGR of 23.7 percent through 2027.[184] For comparison purposes, the CAGR for the global banking, financial services, and insurance (BFSI) industry is *only* 11 percent over a forecasted period 2021–2026.[185]

You may recall me mentioning I consider Mobility as a Service (MaaS) to be the future of transportation. Indeed, there are many reasons that have compelled me to arrive to this argument. As explained earlier, however, MaaS is a relatively newer concept when it comes to transportation trends and, as a result, there are many areas to be explored and questions to be answered before a full MaaS solution deployment.

183 "RTD and Kyyti Group Launch App that Fully Integrates Regular Bus and Rail Services with Flexride Service," Kyyti Group. press release, February 8, 2021, on the Kyyti Group website, accessed February 5, 2021.

184 "Global Mobility As A Service (MaaS) Market 2020-2027 by Service Type, Application, Business Model, Vehicle Type," Research and Markets, GlobeNewswire, December 10, 2020.

185 "BFSI Security Market - Growth, Trends, Covid-19 Impact, and Forecasts (2021-2026)," Mordor Intelligence, 2020.

KEY TAKEAWAYS

- Shared mobility and Mobility as a Service (MaaS) have appeared as innovative mobility solutions to the rapid US urbanism and the associated increase in traffic congestion and environmental pollution.

- The advantages offered by shared mobility solutions are leading to a drop in car ownership rates in the US for the first time in modern history.

- MaaS is a relatively new concept consisting of the movement from personally owned vehicles toward mobility solutions provided by a combination of public and private organizations. MaaS is a digital platform where you can plan, book, and pay for a travel journey all at once.

- Over the past few years, several private companies/startups and public transit entities have been exploring and investing in the development of MaaS solutions. Although most of these MaaS services have an urban focus, MaaS solutions can also be deployed in suburban and rural areas, and at a regional level.

- While MaaS has been popularized mainly in Europe, pilot programs on integrated on-demand mobility have also been developing in the United States.

In the third part of this book, I will take you through the roadmap to what I envision "Mobility 2040" to be—seamless automated and sustainable multi-modal integrated mobility. I will further explain how other mobility trends covered (e.g., electric vehicles, autonomous vehicles) can be leveraged together to enable MaaS and ultimately help reinvent a fragmented US transportation system over the next few decades.

PART 3:

THE ROADMAP TO MOBILITY 2040: SEAMLESS AND SUSTAINABLE MOBILITY

CHAPTER 9:

WHY MOBILITY AS A SERVICE?

———

"Mobility as a Service is the twenty-first century equivalent to the Ford Model T, which gave people the freedom to go wherever and whenever they wanted."[186]

—SAMPO HIETANEN, CEO AND FOUNDER

OF MAAS GLOBAL LTD.

BREAKING THE AUTOMOBILE OWNERSHIP STATUS QUO

The modern history of the US transportation system has taught us it was built around one specific component: the privately owned car. On top of being the norm, cars have become a cultural symbol inherently linked with ideas

---

186 Sampo Hietanen, quoted in Carlton Reid, "Netflix-Of-Transportation App Reduces Car Use And Boosts Bike And Bus Use, Finds MaaS Data Crunch," *Forbes*, March 28, 2019.

of freedom, speed, masculinity, and social status upgrading.[187] As such, culture became gradually dependent upon them, and so entered The Era of the Automobile, defined as a "complex path-dependent non-linear system."[188] The vehicle-controlled scheme has resulted in enormous levels of environmental consumption of resources as one of the principal socio-technical institutions through which modernity is organized.[189,190]

I remember my frustration back in 2007 when I realized I would not be able to [lawfully] drive a car in Ecuador. However, when I moved to the United States after turning eighteen, I was the minimum driving age and could earn the right to get a driver's license of my own. The fear of missing out (FOMO) was real. I thought about my older cousins and the freedom their vehicles had granted them. They could go anywhere, at any time. To me, a driver's license embodied the transition from the dependence of childhood, to the responsibility and freedom of adulthood.

I found it quite interesting to see how for my fifteen-year-old cousin Dennis, it was a far cooler accomplishment to get a

187 Wolfgang Sachs, *For Love of the Automobile: Looking Back into the History of our Desires,* trans. Don Reneau (Los Angeles: University of California Press, 1992).

188 John Urry, "The System of Automobility," *Theory Culture & Society* (October 2004).

189 ohn Urry, *Mobilities* (Cambridge, USA: Polity Press, 2008), quoted in Maxime Audouin et al., "The Development of Mobility-as-a-Service in the Helsinki Metropolitan Area: A Multi-Level Governance Analysis," Research in Transportation Business & Management (Elsevier, 2018).

190 Steffen Böhm et al., "Part One Conceptualizing Automobility: Introduction: Impossibilities of automobility," *Sociological Review* (September 2006).

driver's license than to be able to vote or even be allowed to drink. For him and his high-school classmates, it was the beginning of a brand-new world. Now they could cruise down to meet with their friends and put an end to the ritual of being picked up by their parents.

Interestingly enough, young people's obsessive desire to get driver's licenses has started to die down over the past decade. According to the Federal Highway Administration, the percentage of American sixteen-year-olds with driver's licenses was 25.6 percent in 2018, a drop from 31.1 percent in 2008. If we go back to 1983, a whopping 46.2 percent of sixteen-year-olds had a driver's license in hand.[191] Even though age restrictions vary by state and not all sixteen-year-olds can be license holders, the decline in interest to lawfully drive is evident.

Interesting! Now, how can we explain this trend? While there is more than one valid answer to this question, the technological advancements and new mobility solutions are clearly one of them. They have had a direct impact in the loss of interest in young people's perceptions on getting a driver's license. According to Public Broadcasting Service (PBS), stricter regulation for younger drivers in addition to the availability of ridesharing and ride-hailing apps are why teenagers delay getting their license.[192] Again, think of the convenience and cost savings associated with the Ubers and Lyfts, as explained in the previous chapter of this book.

---

191 Katharina Buchholz, "Americans Get Drivers Licenses Later in Life," Statista, January 7, 2020.

192 Tim Henderson, "Why Many Teens Don't Want to Get a Driver's License," PBS, March 6, 2017.

The outcomes provided by alternate modes of transport other than by a privately owned car are also observed in other parts of the world. Finland, for example, is also seeing a decline in the number of people obtaining a driver's license. As stated by Max Fogdell, Head of the Driving Licenses unit at the Finnish transport and communications agency Traficom, fewer young people are now obtaining a driver's license as they do not perceive a personal car as a must-have.[193]

"The number of young people getting a driving license is declining at a steady pace in all Finnish cities, and the trend is global."

According to driver's license statistics, about 55 percent of eighteen-year-olds who live in cities obtain a driver's license as soon as they can, while in Helsinki just a third of them do so.[194] Fogdell considers that the main reason interest in obtaining a driver's license is declining among young people is that they simply don't need a car when alternative modes of transport are readily available.

"Young people usually get a driving license to travel to work or school. If their place of work or study is easy to access within a reasonable amount of time by public transport, bike, or on foot, then they don't find it necessary."

Moreover, Fogdell refers to the high cost in obtaining a driver's license in Finland—from $1500 to $3600 through a

---

193 "Easy Access by Public Transport and Bike—Young People are Postponing Obtaining a Driving License," *Whim (blog)*, MaaS Global Oy, November 7, 2019.

194 Ibid.

driving school for a Class B car license (the most common license).[195,196]

As such, thanks to the various transportation modes available to the Helsinki population, the Finnish capital appears as one of the cities with the lowest percentages—well below average—of people obtaining a driver's license.

FREEDOM OF MOBILITY

Mobility as a Service (MaaS) takes a step further than traditional shared mobility (e.g., Uber, Lyft) by integrating distinct public and private modes of transportation into a single digital platform. This allows the user to plan, book, and pay all in this same platform (e.g., mobile app) without requiring a license or a skills assessment to do so. By offering a myriad of transportation options, MaaS is a great alternative to the personal car for those who don't own a car, cannot drive, or simply would like to leave their vehicles at home. MaaS is about ensuring freedom of mobility by offering choice.

Sampo Hietanen from Finland is the CEO and founder of MaaS Global Ltd., known for its Whim mobility app. Coming from executive and board positions in civil engineering and the Intelligent Transport Systems (ITS), Hietanen is considered father to the concept of Mobility as a Service (MaaS). In January of 2020, he received the Order of the White Rose of

---

195 Ibid.
196 "Driving in Finland: Licenses, Rules, Vehicles & Tires, Schools," Expat Finland, accessed February 10, 2021.

Finland, awarded by Finland's president Sauli Niinistö, for his outstanding social merit.[197]

In a *Futurebuilders* podcast episode, Hietanen shares his vision on the future of transportation and his experience within the MaaS space.[198] He has been talking about MaaS for over a decade around the world, speaking at more than two thousand events. He says his mission is to explain the potential of MaaS and its ability to disrupt global transportation. Hietanen has been working in the transportation sector his entire life and started his career journey at a Finnish civil engineering company called Destia. Destia was heavily invested in traffic information and was poised to build the world's first short-message service (SMS) based route information system. This job is what would ultimately trigger Hietanen's curiosity on how information systems could disrupt transportation.

"I think it was in 1999 and in that whole traffic information space… I remember I had to give a speech about how technology and digitalization disrupts the whole transportation and what can we expect as changes."

"And I was coming from London when I started comparing what happened in the telecom industry, and what might happen in transportation, and this [MaaS] felt like a huge idea… it was quite an obvious one that the macroeconomics of the whole transportation are such that its bound to be disrupted."

197 "CEO Sampo Hietanen Honored with Finland's Order of the White Rose," *Whim (blog)*, MaaS Global Oy, January 14, 2020.

198 *Futurebuilders Podcast*, "Interview with Sampo Hietanen—Mobility as a Service, the Future of Transportation," April 23, 2019, video, 38:14.

Hietanen says there have been hundreds, if not thousands, of people who have discussed the concept of Mobility as a Service before him. He contends MaaS would have become a reality with or without his contributions. However, he does consider himself to be one of the foremost thought leaders most actively talking about MaaS. He recognizes his extensive history speaking on the topic has taught him the skills to verbalize the concept in such a way that's easy to understand for others.

Hietanen suggests that consumers, mainly younger generations, are prepared to give up their vehicles to rely solely in on-demand shared transportation. They are ready to be served if you provide them a good enough service. He claims younger people, usually those under thirty, would rather postpone having a car for as long as possible or try to get rid of it but nobody else is offering them a good enough alternative.

"I say, 'Hey, wouldn't you like to have all your mobility needs in one single, one-stop shop in a single app? Wouldn't you like to have that with the service guarantee? Wouldn't you like to have access to beautiful cars than the mediocre car that you actually can afford?' And they say, 'yeah of course, that's how it should be."

Hietanen says his company MaaS Global is not asking people to give up their cars. Instead, he is admiring cars as a product and trying to be as good or better. The only way you can disrupt a sector, Hietanen suggests, is by looking at the best thing that already exists and improving upon it. The "father of MaaS" suggests if they really want to disrupt transportation, to do it not by preaching to people or managing them but by bringing to them something effective and efficient. Hietanen believes they can do that.

"If you think of the concept Mobility as a Service, focus on the last word."

Hietanen refers back to the time when he first started thinking about MaaS and how he looked at the analogy of the average revenue per user (ARPU) in the telecom industry, which is the monthly budget people allocate for this need. He says the ARPU did not exist in mobility, and thus he had to try to find it. Luckily, today it's very easy to look it up from consumption.

He explains in European and the Western world, people spend about €30 per month for their mobile phone needs. In mobility, it's ten times that. Therefore, mobility is ten times as big as the telecom disruption was. Hietanen says mobility is actually the second largest expenditure with 76 percent of the value of the market relying on a car. As an asset used about 4 percent of the time, that is a lot. For example, in Tokyo, Japan, over 50 percent of the cars on the island are only used approximately once per month.

"What do I have to do for those people to give me the same money and feel the same comfort? And it boils down to this one thing: if you want to think what MaaS is, it's anywhere, anytime, on a whim. That's what the car brings. That's why the service is called Whim. It's because we ask people what is the essence of this freedom that people deserve."

In one of his company's blog postings, Hietanen explains Whim's business model and how they aim to create a better experience for the user than what a vehicle can offer them.[199] By providing access to a plethora of transportation modes, Whim ensures they have the freedom to choose their commuting journey.

"We are old-fashioned in the sense we buy the parts, then package and brand them to meet our customers' needs, and then charge for the value we create to the customer."

"In practice, this means buying bus, tram, taxi, bike rides, and car rentals beforehand based on our knowledge on how much and how people like to move in a month. Then we assemble these rides into packages that meet different demand profiles and focus on creating an experience that beats owning a car."

During the interview, Hietanen further suggests most people tend to buy cars that although are not necessarily the cheapest, are far from being their dream cars. He says individuals should be able to drive something nicer for those weekend summer trips.

"I like to ask people, 'so how many in the audience have a car?' Almost every hand goes up... And then the second question is, 'how many of you have your dream car?' Outside of Abu Dhabi there are not many hands up."

199 Carlton Reid, "Netflix-Of-Transportation App Reduces Car Use and Boosts Bike And Bus Use, Finds MaaS Data Crunch," *Forbes*, March 28, 2019.

Hietanen thinks everyone in transport wants to emphasize how different their local conditions are, yet the idea of freedom of movement and freedom of mobility is the same everywhere. We want to be assured we can go places, especially in Western free societies. "A car is a symbol of freedom," Hietanen says, and thus MaaS has to make sure that freedom is not jeopardized.

"Car is quite a democratized freedom in that sense but it's still not for everyone. And I think for the whole of society this freedom to be mobile, to be able to go places, to move is quite a profound thing that should be given to everyone."

Besides, people not only deserve to move freely anywhere but should also be able to have a say in what they choose. For instance, Hietanen explains how having a single public operator provide transportation to a low-income community is not really freedom. The more we can have freedom of choice, to be able to pick who we get that freedom of mobility from, the more equitable transportation becomes. Hietanen makes the assertion the transportation is large and complex, thus unlikely to be centrally governed.

In another interview, Hietanen outlines how MaaS Global and the Whim app are creating this freedom for individuals by delivering on a combination of all modes of transportation.[200] This way, they can ensure commuters have every single ride available to them. All MaaS Global does is digitally

---

200 *TheCamp*, "Disruption in Mobility and Why It Will Be Collaborative," October 30, 2020, video, 34:38.

combine all into a subscription. By doing this, Hietanen considers they are providing the same level of freedom to a car.

"All we have to do is digitally combine all of them, mold that into a subscription, and we are at the same level of a car. But all this means is we are able to connect them, and then it scales."

Andy Boenau is the founder of Speakeasy Media, an expert business storyteller and board director of MaaS America—an organization created to advance Mobility as a Service in the United States.

During my interview with Boenau, he also makes mention of Mobility as a Service as freedom of mobility. Before diving into how *cool* MaaS is, Boenau walked me through his transportation engineering career journey before finding his real passion (one of them at least) at the intersection of urban planning and storytelling.

"My parents helped me buy a civil engineering degree. I still didn't know what I wanted to do, so I started in traffic engineering. I grew up in the Virginia suburbs of Washington, DC. I interned for an engineering consulting firm and joined the traffic team because I could visualize it. In Northern Virginia, you're always stuck in traffic."

"I never saw myself in the traditional hierarchy of a consulting firm, working my way up as team leader, division manager, office manager, and regional manager. Before I knew it, it had been five years, ten years, fifteen years, and now twenty-two years working in transportation. I kept asking myself what I want to do when I grow up. I'm still asking."

Boenau details how he had friends who would not understand his career trajectory. He explains how from just asking questions he would learn about traffic analysis, software, or intersection analysis. He says he didn't want his boss to have to do all of his work, and thus would start asking why things were done in a certain way. To Boenau's dismay, the answers were routinely:

"Because that's how it's always been done".

This did not resonate well with Boenau. He says he had been developing a genuine interest in marketing and advertising.

"I have an issue with authority already. But in my work, I was working in kind of two parallel tracks. As I was involved in transportation, my love of advertising and marketing was growing."

Boenau contends he gradually found his "sweet spot" in the transportation industry, at the intersection of mobility planning—how people move around, transport—and the storytelling side of things.

"Not storytelling in the sense of 'let me tell you about the invention of the first automobile,' but spreading messages that persuade people to take some action. Frankly, it just comes down to learning about language and learning what works."

Boenau makes reference to Sampo Hietanen's aforementioned perception and portrayal of MaaS as freedom of mobility. He suggests it's rare for Europeans to talk about "freedom of fill-in-the-blank" the way that Americans do, and refers to it as a cultural difference. Boenau admits, nonetheless, Hietanen's depiction of MaaS as freedom resonates well with him. He claims he had used a similar branding for one of his clients, a sustainable shared mobility company.

"Part of my branding with Gotcha...I helped them into wrapping around this idea of freedom of mobility, so their tagline became we believe mobility is freedom."

Boenau further suggests how MaaS Global's founder talks in the same way as he projects on what MaaS could be, which Boenau considers very smart also from a marketing point of view. As such, alluding to Whim's website and other MaaS operators, Boenau left me with a representation of what a transatlantic travel journey could look like upon a full implementation of Mobility as a Service.

"Imagine a world where you can pull out your phone and boom, boom, boom, 'okay I got my whole trip set. I'm going to head from Helsinki, make a trip to Washington, DC, and then down to Walt Disney World in Florida. While I'm down there, I can get an e-bike because Disney World has e-bikes. I'll add that to my trip.'"

"To be able to do all that would be integrated mobility. Maybe you have different brands—you have Andy's e-bike, you've got an Uber, you've got a car doing something—but it doesn't matter which brands because it's all in one platform and you pay for it seamlessly."

OPTIMIZE YOUR JOURNEY

Hietanen's and Boenau's perceptions on MaaS reminded me of Vincent Valdes's previous discussion in an *ITE Talks Transportation* podcast interview.[201] Valdes worked at the Federal Transit Administration (FTA) on the cutting edge of transportation and transit technology and new practices and demonstrated a peculiar excitement about new technologies coming into play in public transportation. Likewise, he analyzed how companies across distinct modes of transport are gradually collaborating among them and moving toward a Mobility as a Service (MaaS) ecosystem.

When asked about micromobility (e.g., e-scooters, bikes) and how it compares to traditional shared mobility, Valdes suggests the two should not be compared but rather considered as key components of a Mobility as a Service integrated network. The end result, he claims, is to create an optimized, coordinated transportation network designed to meet each individual's transport needs. Just like in many urban areas where walkability and transit scores exist, in Valdes's ideal world a mobility score would be great to have to tangibly measure people's mobility in cities.

"I would love to see a mobility score so you as a traveler, again, knowing that you need to get from point A to point B, can check your mobility score and that might mean you use a bike share if you can, or you might walk, and you have a good way finding technology at your hand."

---

201 Vincent Valdes, "The Impact of Innovation on MaaS/MOD with Vincent Valdes," *ITE Talks Transportations Tracks*, produced by Bernie Wagenblast, podcast, Spreaker, 19:05.

"Or that you use a transit ride, or that you might use a TNC [Transportation Network Company], Uber or Lyft ride, but that you're optimizing your trip based on your conditions, and those conditions might be changing from day to day."

If you ask me, that's exactly what Mobility as a Service consists of. It's about transportation optimization, which can be easily illustrated through our day-to-day events. For instance, I think about the times when I was going to job interviews in the Washington, DC metropolitan area. Although transit (e.g., bus, metro) was always the cheapest option, I would take a taxi or Uber/Lyft to reduce my chances of arriving late, while also refusing to ride my bike to avoid getting sweaty.

In contrast, on days being late was not an issue and I decided to exercise, I opted to ride a bike to go from point A to point B. In both mentioned cases, I was optimizing my transportation choices based on real-time needs and real-time conditions.

A COST-EFFECTIVE AND SUSTAINABLE TRANSPORT ALTERNATIVE

Toward the end of his interview with *Futurebuilders* podcast, Hietanen shares one of my favorite stories where he highlights the economic benefits of MaaS when compared to the private cars associated low utilization rate and efficiency. The "father of MaaS" describes a conversation with a transportation engineer in Dublin (Ireland) who did not see how the Whim mobility app was different to transport systems already in place:[202]

"I remember I was in Dublin, and there was this transport engineer, let's say a skeptical one. He said 'I've gotten Mobility as a Service, I have an annual ticket that gets me into Dublin, and within Dublin, and I have that, and it only costs me €700 a year.'"

"I said 'yeah, that can be MaaS, and you just raised your hand when I asked who has a car. So I said, 'okay, you're a professional so can you roughly tell me what's the percentage of your trips you do with your car. Is it like 40 percent? What is it?'"

"And then he calculated for a while and said 'between 5 percent and 10 percent, probably closer to 5 percent of the trips I make with the car.' 'Fine. You're also a professional so you know the actual cost [ownership] of your car, so can you roughly tell me what the cost of that car is?' It was somewhere between €10,000 and €12,000 a year."

"I said 'well, you realize 95 percent of your trips will cost you €700 a year, and the remaining 5 percent will cost you €10,000–€12,000 a year. Do you see where the business case is?'"

Have I mentioned before how many times I have regretted owning a car? Don't get me wrong, I love my Jeep Cherokee. It's taken me on amazing road-trips over the past few years. However, having to deal with parking difficulties, hectic

---

202 *Futurebuilders Podcast,* "Interview with Sampo Hietanen - Mobility as a Service, the Future of Transportation," April 23, 2019, video, 38:14.

traffic congestion and commutes, and witnessing numerous car crashes, I know owning a car does not come without its costs.

I am not only referring to the actual price (MSRP) we pay for cars, but I think of the involved annual expenses including required insurance premiums, licensing and registration, area permit passes, and personal property taxes. This is in addition to fuel, maintenance, repair costs (e.g., oil changes, tires), and finance charges for those who pursue the financing route. As such, it all adds up. In fact, according to American Automobile Association (AAA) research, the average annual cost of vehicle ownership in the US in 2020 was $9,561, or $796.75 a month, representing the highest cost associated with new vehicle ownership since AAA began tracking expenses in 1950.[203] With an effective MaaS system serving personalized transportation needs, however, this cost could be cut in half.

As I was reflecting on the idea of maintaining freedom of mobility without the need to bear the high costs of car ownership, I remembered I had come across the story of the Joutsiniemis. The Joutsiniemis are a family of five from Helsinki who decided to give up their car after their Citroën broke down in 2019.[204]

The family did not see a reason to purchase a new car. While the parents commute is about seven kilometers each way,

203 "Your Driving Costs, 2020," American Automobile Association (AAA), December 14, 2020.

204 "A Family of Five Gave Up their Car: Life Now Runs Smoothly for Them with Good Cycling Equipment and the Occasional Rental Car," *Whim (blog)*, MaaS Global Oy, September 21, 2020.

they admitted to not minding using public transport or riding a bike. Similarly, the children go to school on foot or by bike. When it comes to enjoying leisure activities and hiking, the family affirms they travel to places by bike, on public transport, and by renting a car through Whim. For the Joutsiniemis, in fact, not having a car represents freedom without compromising on comfort. No car means no need to change tires, take it for service, or pay for spare parts, and when the family rents a car it's guaranteed to be in good condition, new, and clean.

"For me, a car is a commodity I need to get from A to B. I really like the idea of shared ownership."

Additionally, it wasn't a surprise to hear the family mention the €2,500 annual cost of their car rentals was significantly lower than the €7,392 annual cost of car ownership, or €616 per month from Sampo Hietanen's words during a chat with Forbes.com.[205] As a side note, the Joutsiniemis' overall car rental comprised a two-month rental in the summer, a one-month rental at Christmas, and a few weekends in the spring and autumn. Not bad, huh?

As a last note on MaaS—and although already implied—MaaS holds the premise of promoting the use of more sustainable modes of transportation including biking, e-scooters, walking, and public transit. As such, MaaS also appears as a promise for a less carbon dependent transport alternative to the personal car.

---

205 Carlton Reid, "Netflix-Of-Transportation App Reduces Car Use And Boosts Bike And Bus Use, Finds MaaS Data Crunch," *Forbes*, March 28, 2019.

Indeed, an initial study report on the impact of Whim and Mobility as a Service in Helsinki, determined public transportation is at the backbone of MaaS. Similarly, it found Whim users are "multi-modalists" and over 50 percent of the trips were made using sustainable modes of transportation such as cycling, bus, tram, train, and metro.[206]

KEY TAKEAWAYS

- While the US transportation system was built around the privately owned car; the everlasting automobile dependency is progressively shifting toward new available on-demand shared transportation modes.

- As MaaS solutions are starting to appear all over the world, Finland in particular has a well-established sector and provides somewhat of a playbook for what we can expect the rollout of MaaS to look like in the United States. As such, there are worthwhile insights we can glean from the Finnish example.

- When properly implemented, MaaS can provide the same level of freedom as the personal car, allowing people to integrate distinct transportation modes—public and private—and optimize their trips based on real time needs and conditions.

- Mobility as a Service comes off as cost-effective transportation—especially when compared to the elevated car ownership costs—and holds the promise of being a more sustainable transport system alternative to the car.

---

206 Ari Hartikainen et al., "Whimpact," *Ramboll*, May 21, 2019.

CHAPTER 10:

REALIZING MAAS

"Accurate and reliable technology is a crucial piece of the Mobility as a Service model, but the true success of integrated mobility is dependent on the private and public sectors coming together to service the public."[207]

—*BUDDY COLEMAN, CHIEF CUSTOMER OFFICER FOR CLEVER DEVICES*

Mobility as a Service (MaaS) holds the promise for a paradigm in the provision of transportation. Given its nascent stage, however, it will need to undergo a period of development prior to reaching maturity and full implementation. The public and private sectors alike will need to address the following challenges in order for MaaS to become a reliable alternate to current transportation habits in the US.

---

207 *Being Mobility-as-a-Service (MaaS) Ready*, American Public Transportation Association (APTA) (Washington, DC: American Public Transportation, 2019).

RESOLVING THE MAAS IDENTITY CRISIS

There is no official or uniquely accepted definition of MaaS. Instead, transportation researchers, engineers, and planners from academia, industry, and government all have their own definitions of this emerging mobility concept.

During a virtual Autonomy and The Urban Mobility Company workshop in late 2020, Suzanne Hoadley, senior manager and traffic efficiency coordinator at POLIS Network, said one of the reasons MaaS has not fully taken off in Europe is because it does not have a single universally established definition, which leaves room for a multitude of interpretations.[208,209]

"I think MaaS has a bit of an identity crisis. I think depending on who you talk to, MaaS is understood differently. You could talk to five people about MaaS, and they would have a different definition. This was something we identified four years ago, and we see that today, and that's why we see the different models emerging."

Hoadley claims MaaS is still in its pilot phase. She says although it's natural for different MaaS models to emerge, it's not helping that different people conceptualize MaaS in such distinct ways—from on-demand transport, to public transportation, to integrated ticketing, to a third-party platform.

---

208 POLIS is the network of European cities and regions cooperating for innovation transport solutions.

209 *Autonomy*, "How Europe is Moving MaaS -and Vice Versa in partnership with Hogan Lovells," October 27, 2020, video, 1:03:46.

"And just yesterday, I was interviewed by a PhD student who is working on a master thesis she carried out. She looked at one hundred sustainable urban mobility plans around the world, and she said only two of them actually mentioned MaaS. So, I think MaaS is still very much in the pilot phase."

Andy Boenau, founder of Speakeasy Media, talks about MaaS in an Urban Mobility Daily blog. He details how mobility experts and industry outsiders have provided inaccurate depictions of app-based transport offerings. He also suggests complex definitions only create more confusion.[210]

"Industry insiders don't do themselves any favors by giving longwinded, complex descriptions of various mobility-on-demand capabilities."

According to Boenau, the MaaS confusion is further aggravated by software developers and consultants who throw around business and marketing jargon when advertising. He says the simpler the MaaS definition, the better.

"Hear me now and believe me later: building a thriving and sustainable MaaS business begins with a clear and simple description of the problem you're solving."

To contrast, consider the maturation of electric vehicles (EVs), autonomous vehicles (AVs), and even micromobility solutions. While these trends may have emerged prior to

---

210 Andy Boenau, "Why Can't Anyone Build a Thriving and Sustainable MaaS Business?," Urban Mobility, January 13, 2021.

MaaS, truth is they have far less ambiguity in the minds of the general public.

To transportation industry experts: this is a call to exercise simplicity over perfection. Very comprehensive descriptions and detailed meanings of the MaaS concept will only aggravate the MaaS identity crisis. A more mainstream conceptualization of MaaS and storytelling is desired and encouraged. This will help facilitate discussion of MaaS in a way that's easier to understand for all transportation users and stakeholders needed for a MaaS development.

Let's start by facilitating a simple definition of MaaS to the US Federal Transit Administration (FTA). While they have awarded many innovative mobility grants, nearly all related to MaaS, they have not included MaaS on their list of shared mobility options to this day.

MAAS GOVERNANCE: PUSHING TOWARD A COLLABORATIVE ECOSYSTEM

Contrary to the turn of the twenty-first century when original equipment manufacturers (OEMs) were the primary party responsible for serving transport, in coordination with respective transit agencies and authorities, the transportation sector today is comprised of different companies across different transportation modalities.

In the second part of this book, we explored many of the innovative ways automakers and auto-parts providers, technology providers, services providers, and software and data startups are contributing to the transportation sector. From the race toward the deployment of electric vehicles carried out by Tesla, Ford, BMW, Rivian, and General Motors, to the advancements in autonomous driving technologies by Volkswagen, Tesla, Nuro, Toyota, and Aurora. From micromobility solutions pushed by Bird, Lime, and Citi Bike to microtransit systems introduced by Via and TransLoc Inc. From ride-hailing shared mobility services brought by Uber and Lyft to StreetLight Data's transportation planning efforts. Wow—the list is extensive, but certainly not exclusive.

We have also witnessed interesting investments (e.g., Volkswagen in Argo AI), partnerships (e.g., Lyft with Waymo), and acquisitions (e.g., Mighty AI by Uber) in pursuit of a faster and more effective deployment of transportation services and technologies.

The development and delivery of Mobility as a Service (MaaS) needs the involvement and collaboration of a large consortium of diverse stakeholders. But at the same time, this complex ecosystem of numerous "actors" has also contributed to the aforementioned identity crisis MaaS is currently experiencing.

In a recent study, Maria Kamargianni and Melinda Matyas from the Urban Transport and Energy Group at University College London (UCL) explored a holistic approach to designing the MaaS ecosystem. The authors suggested such an ecosystem would be composed of many distinct layers, corresponding to different relationship levels within the MaaS operator. They define the first layer, or "core business" ecosystem, as one comprised of (1) MaaS Provider,

(2) Transport Operators, (3) Data Providers, and (4) Customers/Users.[211] Below is a more detailed description of each one:

- **MaaS Provider/Operator:** Public transport authority or a private firm (e.g., MaaS Global (Whim), BVG (Jelbi)).
- **Transport Operators:** Public and private, such as public bus agencies, private bus/shuttle companies, metro/subway operators, bikesharing companies, car rental companies, carsharing companies, e-scooter companies, etc.
- **Data Providers:** Manage the data exchange among stakeholders by selling their capacity to MaaS providers and giving access to their data via secure application programming interface (APIs) gateways.[212]
- **Customers/Users:** The people booking the trips, traveling, and paying for the service.

Although with a lesser degree of commitment to the MaaS provider, other actors involved in the MaaS ecosystem include technical backend and IT providers, information and communications technology (ICT), insurance companies, regulatory organizations, universities and research institutions, and investors. Kamargianni and Matyas note, however, this list is not exclusive. As the MaaS ecosystem evolves, other stakeholders could also be added.[213]

---

211 Maria Kamargianni et al., *The Business Ecosystem of Mobility as a Service* (Washington, DC: Transportation Research Board (TRB), 2017).

212 An API is a software intermediary that allows two applications to talk to each other. For example, if you're using a third-party travel aggregator such as Kayak or Expedia to book a flight and rent a car, this online travel service interacts with the airline's API and car rental company's API.

213 Ibid.

In short, on the road to a successful implementation MaaS must establish a collaborative environment among an ecosystem comprised of a diverse group of public and private sector stakeholders. Given the public good nature of transportation services, the orchestration of a successful MaaS ecosystem will consist of far more than the traditional stakeholder management we employ in our jobs and businesses. Private and public transportation entities are *both* set to play central roles in the provision of an effective MaaS solution to the population.

UNDERSTANDING MAAS BUSINESS MODEL(S)

Sampo Hietanen—founder and CEO of MaaS Global—speaks on why the disruption in the transportation sector needs to be collaborative. He explains no single mobility player has the ability to provide all transportation modes and services commuters want.[214] Such a business model has too large of a scope for a single entity to manage.

"People really want everything [transportation] from a one-stop shop, but they also want choices of an operator, and they want roaming subscriptions. So how on earth do we do this? And this is where it becomes hard because nobody alone can have all of that. No one has enough supply; no one will ever have enough supply."

Despite this, Hietanen affirms there is a tendency for companies to aim for a unilateral control of the mobility sector.

---

214 *TheCamp*, "Disruption in Mobility and Why it Will Be Collaborative," October 30, 2020, video, 34:38.

"Mobility is big and there seems to be quite a lot of battle of whether I own it, whether I can make the cake myself, whether I can control all of this. Everybody seems to understand that yes, there is a disruption, there will be these operators like us and many others, but can I control it somehow? Can I split the cake before we even have eggs?"

Further, Hietanen contends even if every mobility company (e.g., Uber, Tesla) would have their own physical means of transportation (e.g., taxisharing, bikesharing, scooters), the capital costs would be enormous, and cities would be in chaos. In short, such a system and associated business model would not be viable.

Hietanen concludes the only way we can actually get MaaS going is if there is an open ecosystem where an operator integrating all transportation modes, say Whim, has the same access as its MaaS aggregator competitor counterpart.

"So, sorry, but it will not be an egosystem, it's an ecosystem. That means no single entity will actually dictate this."

Although Hietanen speaks of an open ecosystem as the desired business model for MaaS, truth is the business model question is still up for debate. In an international transport forum discussion paper, Corinne Mulley and John Nelson from the University of Sydney present four possible options for stakeholders in the MaaS ecosystem:[215]

---

215 Corinne Mulley et al., *How Mobility as a Service Impacts Public Transport Business Models,* International Transport Forum Discussion Papers, No. 2020/17 (Paris, France: OECD Publishing, 2020).

- **Walled Gardens:** MaaS providers/operators each have arrangements with the transport operators, not necessarily with all modes, and thus the offers to the user may be different.
- **Public MaaS:** MaaS provider/operator is taken by the public transport authority or public transport operator who makes arrangements with the transport operators.
- **Regulated Utility MaaS:** The public authority regulates the MaaS providers/operators and arranges the transport modes that can be supplied to users. The transport operators have to share their APIs to make possible the interaction among all distinct software in use. This is Sampo Hietanen's above recommended model.
- **Mesh-y MaaS:** Distributed system with automatic transactions happening among users at the center of the ecosystem, it relies on blockchain technology to process transactions occurring between user and transport operator. This model comes with higher uncertainty, as the requisite technology has not yet been proven in this context.

Likewise, a 2019 study mission conducted by the American Public Transportation Association (APTA) in Vienna (Austria), Hamburg (Germany), Hannover (Germany), and Helsinki (Finland), identified three different MaaS "market models":[216]

- **Aggregated public MaaS platform:** "Public MaaS operator takes it all"

---

216 *Being Mobility-as-a-Service (MaaS) Ready*, American Public Transportation Association (APTA) (Washington, DC: American Public Transportation, 2019).

- **Aggregated liberal MaaS market:** "Free market—operators driven" (public and private)
- **Disaggregated public MaaS platform:** "Regulated free market with public enablement"

In short, there is still no perfect, definite answer to what the MaaS business model should look like in the US. As ongoing discussions persist on this topic, Mulley and Nelson have further suggested a business model may significantly vary depending on how the split of public and private responsibilities are governed during the development phases of MaaS.[217]

This last point strongly resonated with me. We should not make heavy weather of establishing a *unique* MaaS business model in the US. We ought not expect a one-size-fits-all kind of MaaS model. The concept of MaaS centers, after all, on the ability to combine different modes of transportation. In urban densely populated cities like Washington, DC, for example, a typical MaaS trip could consist of a scooter ride to the nearest metro station, followed by a metro ride for the rest of the journey. High-occupancy transit would be a primary feature. On the other hand, in rural areas like in previously discussed Johnson County (Kansas) in the Midwest, scooters may not take you too far. Private microtransit solutions like

---

217 G. Smith, *Making Mobility-as-a-Service: Towards Governance Principles and Pathways*, Doctoral Thesis (Gothenburg, Sweden: Chalmers University of Technology, 2020), quoted in Corinne Mulley et al., *How Mobility as a Service Impacts Public Transport Business Models*, International Transport Forum Discussion Papers, No. 2020/17 (Paris, France: OECD Publishing, 2020).

shared shuttles would better serve people's transport needs in in the absence of, say, vastly improved train access in the area.

PUBLIC GOOD AND THE PROFIT MOTIVE: BALANCING INNOVATION WITH REGULATION

When asked about the main catalyst that will start to realize his vision on Mobility as a Service, Sampo Hietanen answers with no hesitation: politics.[218]

"It's going to be politics...the biggest innovations actually happened because of [favorable] regulation."

He further pictures what he would consider a "blast of an innovation": a European-wide MaaS system where all current and future public and private transportation operators across Europe can be utilized under a single API as part of a kind of subscription. Yet, Hietanen remains realistic affirming that without regulatory support these innovations wouldn't go far.

"The sad thing is there are a lot of power games. Well, should we allow these innovations to happen? Are there fears? And can I somehow control all of these changes?... We would have all the tools to show the way in digitalization that we'd be sitting on our hands."

Hietanen's view on the key role of politics and regulatory environment on innovation and businesses, in general,

---

218 TheCamp, "Disruption in Mobility and Why it Will Be Collaborative," October 30, 2020, video, 34:38.

took me back to my MBA classes with professors John Mayo, Marcia Mintz, and Andrea Hugill at the Georgetown McDonough School of Business.

Leveraging its location in Washington, DC, right at the intersection of business and public policy, Georgetown's MBA program offered a certificate in nonmarket strategy. Such program aimed to promote a deeper understanding of the ways in which business success is shaped by regulatory, legal, political, and cultural forces beyond the market, including the critical relationships and interactions among firms, government, and the public.

All of this emphasizes how companies from distinct sectors with exemplified market success (e.g., Airbnb, Uber, Wells Fargo, IKEA, Walmart) can struggle and even fail when confronted by forces in the nonmarket environment. For this reason, they need to recognize and understand the "market imperfections", and ultimately adopt a strategy to mitigate challenges for business success. Experts in nonmarket strategy David Bach and David Bruce Allen explain:[219]

"The challenge for CEOs and their leadership teams is one of simultaneous separation and integration. To manage successfully beyond the market, executives must recognize the important differences between the company's market and nonmarket environments but then take an integrated, coherent, and strategic approach to both arenas."

---

219 David Bach et al., "What Every CEO Needs to Know About Nonmarket Strategy," *Research Feature*, MIT Sloan Management Review, April 1, 2010.

With the above-mentioned statement, I have no intentions of suggesting *the* nonmarket strategy for MaaS Global, Jelbi, and so on to "work around" the regulatory environment whatsoever. Instead, I aim to emphasize the importance of an effective collaboration between the private and public sectors, and the role of government in the development of the transportation sector which, after all, has major socioeconomic implications.

Policy makers and citizens: national conversations are needed on how we must respond to the myriad of challenges transportation faces. In light of our country's transportation system improvement, discussions on how to enable mobility technologies and solutions are desired.

PROMOTING INNOVATION WITH A PURPOSE: TRANSPORTATION RESEARCH BOARD (TRB)

Neil Pedersen shared with me his vast and rich career journey within the transportation industry, and his belief in the need for government and the private transport sector to collaborate together. Pedersen is the executive director at the Transportation Research Board (TRB) of the National Academies, Engineering, and Medicine. Prior to joining TRBs ongoing Strategic Highway Research Program (SHRP), in 2012 he worked for the Maryland Department of Transportation's (MDOT) State Highway Administration (SHA).

The Transportation Research Board (TRB) is a division of the National Academy of Sciences, Engineering, and Medicine, which serves as an independent adviser to the President of the United States, the Congress, and other federal agencies on scientific and technical questions of national importance. TRB promotes innovation and progress in transportation through research in an objective and interdisciplinary setting.[220]

During our interview, Pedersen explained how, given his planning background, he tends to think about the future and toward the future. For this reason, he claims one of the areas he has been very interested in is having TRB very involved in trying to stay at the leading edge of the use of technology for mobility enhancement. Pedersen manifests that very early on, TRB was getting into autonomous vehicles and in his view, while vehicles are not becoming autonomous yet, they're getting closer to it. In the midst of all of this, he says transportation network companies (TNC) (e.g., Uber) and shared mobility services started, followed by micromobility.

Pedersen highlights the role TRB plays in assisting local and state governments with how they should consider and approach new mobility developments. He says we need to be thinking about what needs to be done, whether it's from a statutory, regulatory, or policy standpoint.

"And I think we, TRB, really can provide some of the greatest value. We're not going to be on the technology side. We are not going to be on the vehicle development side or in the software technology side but thinking more about how we serve especially state and local governments as they have to be thinking about this."

---

220 "About | Transportation Research Board," The National Academies of Sciences, Engineering, and Medicine, accessed October 8, 2020.

*"What are the policy issues that need to be thought about...
from an economic development standpoint, from a safety
standpoint?"*

*"So, all these are areas that were spending a fair amount of time
on, both doing research as well as convening activities where
we bring a lot of people together to talk about these issues."*

A platform like TRB for discussions among professionals
in the private and public transportation sector is absolutely
necessary for the implementation of mobility solutions in
the country. Moreover, I coincide with the need to promote
innovation with a purpose, ensuring broader transport societal goals are pursued and achieved. We need to consider
sustainability, congestion, safety, and accessibility related
issues associated with the proposed technologies.

Pedersen also suggests what they (TRB) are trying to do is
research and have discussions about the policy issues related
to emerging technologies. He refers to the inherent conflict
between the public interest and the profit motive and believes
it's not advisable to completely deregulate the private mobility sector given their narrow profit interests.

*"In one policy approach, you could just let the free market do
whatever it wants and have no regulation whatsoever. I would
argue that is probably not a very wise choice because private
companies will end up doing whatever maximizes their profit,
as opposed to really trying to achieve some broader goals."*

He refers to the importance of assuring accessibility for the
economically disadvantaged and the disabled community.

Equity considerations are particularly important, Pedersen says, because the free market will tend to provide services where profits will be maximized, rather than ensuring equitable access, especially to lower income areas.

Pedersen also makes mention of ride-hailing ridesharing services and explores services like Uber or Lyft which require a smartphone and a credit/debit card to use. He questions what these companies can do for those who don't have either of them.

"There is still a significant portion of the population that does not own a credit or debit card."

To Pedersen's point on transportation inequity, Washington, DC-based think tank Pew Research Center says although the vast majority of the US population—96 percent—own a cellphone of some kind in 2020, the share of people owning smartphones is 81 percent.[221] In theory, 1 in 5 people would not have access to ride-hailing services.

Yet, speaking to the innovative spirit of ride-hailing companies and to their efforts to serve all the population, Uber and Lyft, for example, now allow users to use mobile payment methods like PayPal and Venmo. Additionally, Uber customers can also pay cash directly to their drivers. As such, credit/debit cards or smartphones are no longer necessarily deal breakers. So, there you go, innovative companies innovating to meet inclusive, societal goals.

---

221 "Mobile Fact Sheet," Pew Research Center, June 12, 2019.

Further, Pedersen says mobility companies tend to focus on the areas where they can make the most money rather than ensuring most people can get access to transportation services.

"So you don't have any problem at all finding Uber or Lyft in downtown Washington, DC or Georgetown but go east to the Anacostia River and it's a lot harder. So, what do you need to do from a policy perspective is make sure low-income areas are being served as well?"

To the above point, I consider quite normal the desire for private companies to focus their services on those areas where they can be profitable. They need to ensure the sustainability of their business. After all, the private industry has a profit motive, and it should not be demonized for it. Having said that, I strongly agree with the idea of deploying more inclusive transport services available to all the population, from affluent areas to low-income neighborhoods. As it was explained in earlier chapters, transportation inaccessibility is a recurring problem and we need to address it.

To this point, in the sixth chapter I referred to how we are already seeing efforts by private micromobility companies to equally serve the population. In Washington, DC, Capital Bikeshare has deployed an effective infrastructure network of docks across all 8 DC wards, providing accessible transportation resources for everyone. Similarly, per the 2021 DC Department of Transportation (DDOT) terms and conditions for the public right-of-way occupancy permit, dockless scooters are made available to all people in the city and maintained in each ward.[222]

Policy makers and private mobility companies: a balance between innovation and regulation is certainly possible in light of the development of our nation's transportation system, serving all people's transport needs.

Likewise, Pedersen recognizes and reflects on how some of the more progressive communities are making significant efforts to offer multi-modal transportation. There is the intention to try to get public transit agencies, and other mobility and technology services to complement rather than compete with each other. As the DC example above, he hopes that we will see more services in more places providing that first last-mile service (e.g., scooters, bikes) to transit stations. The goal being to help provide service to areas that don't have adequate bus service, or during evenings, nights, or weekends when you really can't justify transit service from an economic standpoint.

"Maybe you try to think of ways you can provide subsidized TNC service as an alternative, or in low density areas where right now we're really paying an arm and a leg for a transit service. Maybe there are better alternatives, like the Via type like microbus services, or advanced services, and those kinds of things as well."

---

222 "Final Dockless Scooter Terms and Conditions," Government of the District of Columbia Department of Transportation, 2021.

"So really try to think about how all of these services can oper-ate as a multi-modal system, as opposed to a whole set of individual systems that are not just competing with each other, but really not complementing each other."

Pedersen's argument on the need to deploy a multi-modal transport system based on a collaborative effort speaks directly to the notion of Mobility as a Service. As mentioned earlier, MaaS is about establishing an ecosystem (not an "ego-system") of public and private mobility stakeholders working together. Again, collective action is the key to implementing seamless mobility for all the population.

PUBLIC SECTOR VIEWS (APTA)

In my interview with Nicole DuPuis, director of technol-ogy and innovation at the American Public Transportation Association (APTA), she spoke about some of the historical and existing tensions between the private and public sec-tors, especially when it comes to the deployment of mobility innovations. She says one of the roles of transit is to reinforce environmental justice equity in the mobility ecosystem, and hence sees public transit as the backbone of the transporta-tion system.

"...We must inquire about the role transit plays in the mobil-ity system of the future, especially with all of these new and exciting things such as autonomous vehicle technology, Mobil-ity as a Service, micromobility, and transportation network companies. We still see transit as the backbone of the mobility ecosystem because of the important role it plays in equitably connecting all people."

I do agree with DuPuis on the importance of having a robust public transit network. More so, an improved public transportation system with shorter headways and enhanced accessibility is greatly desired by our communities. We must keep pressuring our elected public officials to invest in and radically transform and improve public transportation. Such improvement would actually accelerate the successful implementation of a MaaS solution.

I believe, nonetheless, rather than seeing transit as an absolute hard dependency for functional MaaS, we need to remain flexible as what mobility solutions can be leveraged more effectively to serve people's diverse transport needs.

DuPuis also speaks to the tremendous opportunities public-private partnerships represent in mobility and refers to the case of the Dallas (Texas) area where transit agencies have partnered with TNCs for first last-mile services.

"I think there is an unbelievable amount of untapped potential in those kinds of public-private partnerships."

She thinks, however, when it comes to public-private partnerships, the public sector organizations don't always come out on the better end of these deals. Tensions emerge when companies won't truly be egalitarian with their public sector partners and won't share data, for example. According to DuPuis, the public sector is often wary of these arrangements because history has shown they don't always emerge as an equal partner.

"In addressing issues of data sharing, I've heard suggestions that to circumvent some of these tensions there should be a third party that collects and manages ridership data… I think we still have a lot to work out on the public policy front if we are really going to pursue some of these public-private partnerships between private sector mobility platforms and public transit agencies."

When it comes to what to share, when, and how, she hasn't seen a truly egalitarian arrangement. However, she "absolutely" sees potential for a truly robust mobility network in the US, comprised of public and private sector mobility providers.

"There is going to have to be a sense of mission-driven goodwill from the private sector and a divorce from the notion of everything being market driven. Because the reality is transportation cannot be solely market driven. It is a human right and there is a huge social safety net component to transit… You have two parties trying to provide a service, but they have very different value systems."

Reflecting on DuPuis's words, I believe that when implemented well, public-private partnerships have proved to be extremely powerful.

Just think of the collaborations between NASA and Elon Musk's SpaceX private aerospace manufacturer and space transportation services company. Two entities with, in theory, very different "value systems."

In May 2020, SpaceX historic final demonstration mission—a two-month test flight—resulted in NASA certifying SpaceX's system to carry astronauts. Six months later, NASA-SpaceX's Crew-1 mission made history for becoming the first commercial crew rotation to the International Space Station.[223] In Phil McAlister's words, NASA director of commercial spaceflight development:[224]

"With this milestone, NASA and SpaceX have changed the historical arc of human space transportation."

Such inspiring groundbreaking advancements in transportation space were made possible as a direct result of solid private-public collaboration. Aside from the space travel milestone such partnership has achieved, it has led to substantial economic savings for NASA. The agency said it has saved $20–$30 billion over the past decade by funding Boeing and SpaceX to build rocket technology.[225]

If the aerospace industry is succeeding through public-private partnerships, you better believe the transportation sector can. Like the motto says, "united we stand, divided we fall."

---

223 Michael Sheetz, "SpaceX is about to Launch its First Full NASA Crew to the Space Station: Heres What You Should Know," *CNBC,* last modified November 15, 2020.

224 Ibid.

225 Michael Sheetz, "NASA Estimates Having SpaceX and Boeing Build Spacecraft for Astronauts Saved $20 Billion to $30 Billion," *CNBC,* last modified May 13, 2020.

REGULATORY ENVIRONMENT: EUROPE VERSUS US

Pedersen suggests there are major differences between the transportation systems in Europe and the United States. In many European countries, they try to provide carsharing, bikesharing, or micromobility sharing services at transit stations. They provide a single app with all the different services and you can choose what you want from it.

"That's the difference between many European cities and American cities. Uber is present in virtually every American city and so is Lyft."

One of the principal differences between Europe and most American cities is they have much more intense development. The cities tend to not be as spread out as American cities and therefore they tend to have stronger transit systems with far more frequent services.

Pedersen says the key to MaaS being successful is to have a very good transit backbone system. For instance, he explains if you're only going to be catching a bus at half-hour headways, it's not going to work. In most of the European cities the bus services have five-minute headways, at least during the peak periods. Pedersen thinks that is one thing that will be a challenge from an American standpoint.

"In Europe if you miss a bus, you're not going to have to wait more than five minutes for another one."

"And when you're waiting for an Uber, if it shows up a little bit late to get you to that bus stop you're going to with half-hour headways, and you've just missed the bus, what are you going

to do? You're not going to wait for the next bus; you're going to stay in your Uber and pay whatever you have to pay to where you have to go."

Pedersen believes there are challenges to finding the right balance between innovation and regulation in the implementation of mobility solutions, including Mobility as a Service. He has suggested the United States faces an additional challenge compared to Europe, since many American cities lack a robust public transportation system at this time.

Upon conducting a study mission of MaaS developments in Europe through the American Public Transit Association (APTA), Rob Gannon, general manager of King County Metro in Washington State, also speaks to the importance of public transportation ecosystem.[226]

"Public transportation is at the core of any viable Mobility as a Service strategy and requires a new way of thinking and unified approach among public and private transportation providers to deliver shared outcomes and benefits."

Again, allow me to disagree, partially at least. Yes, a robust public transportation system would greatly enhance the entire mobility ecosystem. It would accelerate the implementation of a robust MaaS ecosystem. Yet, I do not see it as a deal breaker toward a MaaS implementation.

---

226 *Being Mobility-as-a-Service (MaaS) Ready*, American Public Transportation Association (APTA) (Washington, DC: American Public Transportation, 2019).

In Europe, public transit is unequivocally the backbone of transportation. However, as Pedersen explains above, European cities are generally not so spread out and thus urban sprawl is not a real concern as it is in the US. We must therefore respond to our own history and built environment reality.

Andy Boenau, board director of MaaS America—an organization founded to advance Mobility as a Service in the US—also makes reference to differences in the regulatory environment between Europe and the US. He refers to a more conservative and skeptical position adopted by local groups when assessing the potential implementation of transportation innovations in the United States:[227]

"Planners and engineers are very conservative people just generally speaking, so there is in any aspect, whether it's a new type of pavement or a new way of designing a bridge, this tendency to wait and see. I want to see. I want to see the proof of concept, and to them a proof of concept is not the same as a startup would say, like, 'okay, we tried this thing for a month or for ninety days.'"

"Civil engineers and transport planners want you to show proof of concept for eight years running and thirty projects, and it just doesn't work well with something like integrated mobility platforms. The status quo insists we wait and see."

To Boenau's above point on what he calls the "wait and see" by US transportation officials, I now reflect on how naive I

---

227 *Autonomy*, "How Europe is Moving MaaS -and Vice Versa in partnership with Hogan Lovells," October 27, 2020, video, 1:03:46.

was to think new changes could be easily and quickly implemented in the transportation design code. I'm referring to my experience at the TRB's 2013 Annual Meeting conference.

My group and I presented the results of an experimental investigation on concrete bridge decks. The Virginia Department of Transportation (VDOT) was progressively moving toward the implementation of corrosion-resistant reinforcing (CRR) bars into concrete bridge decks. The objective of our study consisted of investigating what the design and detailing criteria should be for these bars to be used. Upon leaving the conference and publishing our findings in TRBs journal, I thought it would be a matter of one year for our "innovative" bridge deck design recommendations to be implemented into the US transportation code. Of course, this didn't happen, and to Boenau's point, VDOT wanted to see our proof of concept running in a series of further long-lasting experimental research projects.

BRIDGING THE US TRANSPORTATION INFRASTRUCTURE GAPS

Improvements in the US transportation infrastructure would accelerate the deployment of a more functional MaaS solution. As it's been explained, MaaS is about ensuring freedom of mobility by offering choice. It consists of promoting the use of multi-modal transportation.

With the rising concern for balancing the needs for all roadway users, and the expansion of ride-hailing companies (e.g., Uber), and online shopping and associated deliveries (e.g., Amazon Prime), demand for curbside pickups, drop-offs and dwell times is rapidly increasing.[228] More specifically, curbside users today include drivers (TNC and non-TNCs), pedestrians, bicycles/scooters and infrastructure, transit and transit infrastructure, parked electric vehicles and EV charging, couriers and delivery vehicles, shuttles, and ADA access, among others.

Recognizing the highly competitive urban real estate the space between street and sidewalks represent, distinct entities are putting efforts into curbside management. The goal is to optimize and allocate curb spaces to maximize mobility and access for the myriad curb demands. The National Association of City Transportation Officials (NACTO) and the Institute of Transportation Engineers (ITE) are examples of organizations introducing regulatory, operational, and technology guidelines for effective curbside management.[229,230] Moreover, through a curbside management solution and leveraging emerging technologies like IoT, City Tech—a Chicago-based urban solutions accelerator—is mapping curbside activity and identifying areas to produce efficiencies. In partnership with technology and analytics companies Bosch and HERE Technologies, and data and modeling firms Stantec, SpotHero, Teralytics, and Carrier Direct, City Tech's initiative aims to make city curbs dynamic and valuable assets in urban transportation systems.[231]

---

228 "The Alliance," MaaS Alliance, accessed October 8, 2020.

229 "Curb Appeal: Curbside Management Strategies for Improving Transit Reliability," National Association of City Transportation Officials (NACTO), accessed February 21, 2021.

230 "Curbside Management Resources," Institute of Transportation Engineers (ITE), accessed February 21, 2021.

231 "City Tech Launches New Solution to Address Urban Curbside Chaos," City Tech Collaborative, January 29, 2020.

When thinking about a MaaS deployment, I particularly think of micromobility options as important components for such MaaS transport ecosystem. As such, curbside management in favor of the implementation of safe dedicated bike/scooter infrastructure is desirable. In many cities today, individuals do not feel safe in their rideables when having to share the road with cars going at 45 mph. Similarly, riding on the sidewalks creates hazardous conditions for micromobility drivers and pedestrians. As a result, many opt for an Uber trip or car rental as alternate functional transport options. Because the established infrastructure only allows for one type of safe transportation—the car.

Policy officials and transportation planners: the implementation of safe street and sidewalk infrastructure cannot be ignored. Fortunately, efforts are already being put into this, and companies like StreetLight Data are working on infrastructure prioritization—determining the best locations for bike lanes and pedestrian crossings—by identifying walking and biking activity patterns.[232]

Moreover, as people in cities will start gradually shifting away from private cars toward other transportation modes, by extension the profitability of things like parking garages will start to decline. As such, as MaaS will unlock many of these spaces for other purposes (e.g., parks, apartments, football/soccer pitches), it will be a no-brainer to restructure the real estate space in our cities to a certain extent.

---

232 "Infrastructure Prioritization," StreetLight Data, accessed February 25, 2021.

Additionally, reflecting on my unique visit to the "City of Bridges," Pittsburgh in 2015, it's inevitable for me to envision 2040 with at least a few of the city's numerous bridges repurposed for mass transit bridges as the balance of commuter volume shifts away from cars toward buses and shuttles.

Moreover, while Elon Musk has dominated the headlines for the aforementioned SpaceX rocket launches and Tesla's electric and autonomous vehicle advancements, his transportation venture The Boring Company has also been making remarkable progress. Musk's vision is to solve urban transportation problems like traffic congestion and vehicular air pollution by building tunnels under cities while deploying self-driving electric cars. We are already seeing promising developments across US cities:

In late 2020, Musk announced the near-completion of the first operational tunnel under Las Vegas for self-driving cars.[233] This tunnel system holds the promise to transport passengers in shared shuttle AVs between any two destinations in the city within minutes. Similarly, the San Bernardino County Transportation Authority (SBCTA) in California started negotiations with The Boring Company in early 2021 for a proposed underground transit loop system connecting the Ontario Airport and the Rancho Cucamonga Metrolink Station.[234] Chairman of the San Bernardino County Board of Supervisors Curt Hagman underscored the myriad benefits of the said project:

233 Sissi Cao, "Elon Musks Tunnel Under Las Vegas for Self-Driving Cars Is Almost Complete," *Observer*, September 16, 2020.
234 Maria Merano, "The Boring Company Talks for Ontario Loop Project Begins in San Bernardino County," *Teslarati*, February 4, 2021.

"The Ontario Airport Loop project represents an innovative, cost-effective, and sustainable approach in meeting the mobility needs of one of the most robust population and economic centers in the United States."

Other large-scale projects in the proposal stage include Fort Lauderdale and Chicago.[235]

In short, the general trend of retrofitting tunnels under cities could end up bridging some of the urban infrastructure gap, if momentum in this space continues, thus bolstering the MaaS ecosystem. It's promising to see active interest from California, Florida, and other places.

ADVANCING MAAS VISION AND AGENDA

It's clear technological, governance, and regulatory areas need to be fully explored and understood to successfully implement MaaS in the US. Having said that, there are a number of organizations heavily contributing to advancing the vision and agenda of MaaS today—both in Europe and in the United States—which make me optimistic about a MaaS definite implementation over the next years.

This is the case of MaaS Alliance, a public-private partnership creating the foundations for a common approach to MaaS in Europe.[236] Considering the numerous MaaS initiatives starting up all over Europe, MaaS Alliance facilitates cooperation through a shared work program involving transport

---

235 Joey Klender, "The Boring Co.'s Projects Are Making Transit Departments Rethink Above-Ground Travel," Teslarati, February 3, 2021.
236 "The Alliance," MaaS Alliance, accessed October 8, 2020.

operators, service providers, and users. MaaS Alliance encourages new ideas in light of a MaaS ecosystem development across Europe and worldwide. The organization is made up of three "working groups" addressing issues related to user needs, regulatory challenges, governance and business models, technology, and standardization.[237]

MaaS Alliance is governed by a board of directors comprised of key stakeholders from the public and private sectors, and from distinct countries. It is driven by its members and partners, including public transport authorities (e.g., Royal Dutch Transport, Barcelona Metropolitan Transport Authority), local municipalities (City of Helsinki, City of Vienna), educational institutions (e.g., Aalto University, University of Surrey), and private mobility operators (e.g., MaaS Global, Uber), among many other organizations and institutions.

In the US, MaaS America is a forum founded to advance a vision of MaaS that reflects the American form of mobility.[238] It promotes the development of practical, innovative, and integrated MaaS solutions by connecting public and private sectors to exchange knowledge and ideas. MaaS America has a technology committee addressing the technological state of the practice of MaaS, and a policy committee serving in an advisory capacity to the board.[239]

Coming from the Virginia Tech College of Architecture Urban Studies and the Virginia Army National Guard, Tim

---

237 "Working Together," MaaS Alliance, accessed October 8, 2020.
238 "MaaS America: Advancing the New Mindset of Mobility in America," MaaS America, accessed October 8, 2020.
239 "What We Do," MaaS America, accessed October 8, 2020.

McGuckin, founder and head of MaaS America, shared with me in an interview his motivation of forming the organization after learning about the concept of MaaS in Europe:

"I first learned to the term MaaS in Europe. I was invited to make a presentation for a company I was working with called A-to-Be. I was the US CEO, and it was a Portuguese company that did a lot of integrated mobility services, tolling, parking, ferry bus integration…so in a way it was MaaS before MaaS was kind of termed."

"So I went over there [Portugal] to a workshop… I had two days of just being exposed during session after session on Mobility as a Service and it was so exciting. And then I saw this was something I think we needed to promote in the US."

McGuckin suggests, nonetheless, he felt the history, economy, built environment, geography, and governance elements of Europe are different than in the United States. He felt to get some traction, they needed to have their own version of MaaS.

"I felt there is room for an organization that advanced MaaS in a way that more practically reflects the built environment of the United States."

More specifically, McGuckin explained how given the higher US car dependency compared to Europe, the premise of asking people to give up on their cars may not easily resonate with the American population. Referring in particular to multi-car family owners living in suburban and rural areas where a car becomes more necessary, McGuckin suggests they could start being liberated from the realities of car ownership.

"By population, over half of America still lives in areas that do not have excellent transit connections and you are required to have a car. So, I thought MaaS should not begin advancing in a way that eliminates the car but would help you make the decision as a family perhaps to own one less car. American families can go from four cars, and there are those families, to three to two, or two to one."

He speaks with personal experience going from three cars to only one.

"I have one car. I had three and one was sitting around, and I thought, 'why do I have this?' So, I got down to two. Then I just thought, 'you know, I'll try to live the life' and I got down to one car. I live in an area that is mildly walkable. I can walk to a movie, to the supermarket, or to the doctors or chiropractor."

Another way MaaS America is advancing the vision of MaaS, McGuckin explains in an ARC's *Smart Cities* podcast, is by at least equating the value of the private sector with the one of the public sectors.[240]

"We need people who want to invent and innovate, and that means they want to do something to make money. And there

240 Tim McGuckin, "Mobility-as-a-Service An Interview with MaaS Americas Tim McGuckin," October 28, 2019, in *The Smart City Podcast*, produced by Eduard Fidler, podcast, *ARC Advisory Group*, 1:18:33.

is nothing wrong with that. So, we looked at what is the new paradigm for transportation under this umbrella called MaaS."

On a more personal note, coming from a civil engineering background and always in sight of the American Society of Civil Engineers (ASCE) developments, I've been serving as the vice chair of the ASCE Technical Committee on Mobility Innovation (TCMI) since its formation. The committee consists of a forum for bridging strengths in academia, industry, and government to promote MaaS technology, platforms, regulations, policies, and operating systems.

Setting the standard for how MaaS projects are evaluated, equity implications of emerging mobility options, and MaaS' role in reviving public transportation are a few topics we have brought to the table for discussion among transportation experts from the US, Europe, Asia, and Australia.

Other MaaS-promoting stakeholders in the United States include the American Public Transportation Association (APTA) Research and Technology Committee, the American Association of State Highway and Transportation Officials (AASHTO), and the Transportation Research Board as previously narrated by its executive director Neil Pedersen.

While the challenges toward a full implementation of MaaS are certainly there, a number of organizations in the US and Europe are taking important steps to drive MaaS vision and agenda considering the technology and regulatory environments.

This leads me to be convinced MaaS will successfully see its full implementation in the US over the next years, helping us achieve seamless multi-modal integrated mobility.

KEY TAKEAWAYS

- Mobility as a Service (MaaS) holds the promise for a paradigm in the provision of transportation. Given its early stage, however, it will need to undergo a period of development prior to reaching maturity and full implementation. Key areas need to be fully explored and questions answered on the road to a definite MaaS solution deployment.

- Hurdles a MaaS implementation in the US must overcome are a MaaS identity crisis, the need to establish a large collaborative ecosystem composed of private and public sector stakeholders, understanding and choosing a viable MaaS business model(s), finding the right balance between innovation and regulation, and bridging the US transportation infrastructure gaps.

- Despite the technological, governance, and regulatory barriers MaaS foresees, there are numerous private-public sector partnerships promoting and advancing the vision, agenda, and development of MaaS in the US.

CHAPTER 11:

ELECTRIFYING AND AUTOMATING MAAS

"Frictionless, automated, personalized travel on demand—that's the dream of the future of mobility."[241]

—*DELOITTE SERIES ON THE FUTURE OF MOBILITY*™

As discussed in previous chapters, the future of transportation lies in Mobility as a Service. Digital platforms through which people will be able to plan, book, and pay for their trips all at once. It will be a system where users will have access to a menu of private, public, and shared transport modes, including micromobility and microtransit options. So, the natural question is how can the advancements in autonomous and electric vehicles then be leveraged in a MaaS ecosystem? I envision the US transportation system as autonomous and sustainable multi-modal integrated mobility. This system

---

241 Scott Corwin et al., "The Future of Mobility: Whats Next?", Deloitte Insights, Deloitte, September 14, 2016.

will be Mobility as a Service enabled by autonomy and electrification—call it AeMaaS.

Considering the technological and regulatory barriers, a deployment at scale of Level 5 fully autonomous vehicles will take time. It will not happen overnight. It won't even be a matter of a few years. Despite their increasing adoption, it will take time just to boost the number of electric vehicles on US roads and highways. Mobility as a Service (MaaS) is still in its early stages, and I foresee a development period before it reaches maturity and is in full implementation.

Nevertheless, I am still convinced rapid advancements in cutting-edge technologies such as artificial intelligence (AI), big data, the internet of things (IoT), and 5G connectivity will have great impacts on the road ahead. With effective collaborations between incumbents and disruptors private and public sector mobility stakeholders, in addition to a more progressive transport legislation and regulatory environment, we can achieve this new vision on mobility by 2040. The reimagined US transportation system that urban, suburban, and rural communities all yearn for and deserve will be a seamless, sustainable, and inclusive mobility ecosystem serving the transportation needs of the entire US population.

According to a collaborative study between Capgemini Invent and Autonomy, by 2030, revenues already generated by autonomous vehicles, electric vehicles, and MaaS are expected to increase tenfold.[242]

---

242 Mehdi Essaidi et al., "The Future of Mobility as a Service (MaaS): Which model of MaaS Will Win Through?," Capgemini Invent, Capgemini and Autonomy, 2020.

ENABLING MAAS THROUGH AUTOMATION

In an interview on the *Futurebuilders* podcast, Sampo Hietanen refers to the integration of autonomous vehicle (AV) services as "the ultimate enabler" to the MaaS ecosystem.[243] Hietanen says AVs will hit the roads as car ownership significantly decreases. However, he believes the deployment of AVs should be under a shared system rather than a private personal car type of use.

"That's when you're not going to own an automated vehicle, that's for sure, or maybe as an investment but not really. Why would you let it lie over there when you can put it to use? And the fleets needed are about 5 percent of the cars we now have on the roads."

Hietanen says the AV simulation would have a robo-taxi pick a person up anytime from their home and take them all the way to their destination. Although it sounds like an extremely convenient road travel journey, he suggests such a situation would ultimately lead to over 100 percent more traffic.

"I would not like to be in that city; it would be perfect for the first 2 percent. We've got an excellent service, good price, but then the end effects would be horrible."

I could not agree more. Now, with thousands of AVs flooding the roads, could such a situation be mitigated by simply stopping or banning those automated cars? Probably not. Instead, to please the people and offer them the services they

243 *Futurebuilders Podcast*, "Interview with Sampo Hietanen—Mobility as a Service, the Future of Transportation," April 23, 2019, video, 38:14.

want, politicians may suggest a road expansion as an *easy* solution. I get goosebumps just thinking about that possibility. It takes me back to those times of the unorganized, unfocused, and dysfunctional Interstate Highway System expansion. In other words, back to square one. Recalling British statesman Winston Churchill's words, paraphrasing philosopher George Santayana:

"Those who fail to learn from history are doomed to repeat it."[244]

This unappealing first AV deployment scenario also reminded me of my interview with Neil Pedersen. Pedersen illustrates in more detail what the negative implications of a personal AV ownership and trip would be.

"If someone owns a driverless vehicle and they have that vehicle drive them to work, and they just send it back home again, or they take it on a shopping trip and they just have the vehicle driving around while they're shopping... It's going to very quickly overwhelm the system from a congestion standpoint."

Pedersen is right. Autonomous technologies could do more harm than good if they lead to significantly more automobile trips being made. Is this a reason to panic? Absolutely not! Fortunately, there is a much better AV deployment scenario on which the vehicles would be shared and serving as connectors to mass transit. This way, traffic congestion would significantly decrease instead. Hietanen also refers to this:[245]

244 Winston Churchill, quoted in "History Repeating," College of Liberal Arts and Human Sciences, Virginia Tech, accessed January 15, 2021.

245 *Futurebuilders Podcast,* "Interview with Sampo Hietanen—Mobility as a Service, the Future of Transportation," April 23, 2019, video, 38:14.

"If those AVs will be such that they will be more shared, they will pick you up, take you to a station where you have more mass transit, and work in a more productive way...the difference between those scenarios is huge."

Not surprisingly, Neil Pedersen's opinion on how autonomous vehicles should be deployed coincides with the above second scenario.

"As we move more and more toward autonomy, the key to being able to manage the system, and particularly manage congestion, is getting a very large portion of travelers to actually share vehicles and use shared vehicles as opposed to continue to own."

Paula Bejarano shared with me her views on the future deployment of AVs. She suggests, not only from a congestion but also economic standpoint, AVs should be deployed under a fleet rather than individualistic model.

"Creating the new context of the future with a [AV] fleet model rather than an old model of vehicles...autonomous vehicles in the future will most likely be mainly owned by fleet operators, not individuals."

Todd Litman, founder and executive director at the Victoria Transport Policy Institute, presents the advantages and disadvantages of distinct models of AV transport provision. He notes for a personal AV model, the external costs including congestion, facilities, crashes, and pollution would be high as increasing total vehicle travel tends to increase external costs.[246] On the other hand, assuming a good application of microtransit/shuttle type AVs, this model would present lower external costs.

A recent McKinsey analysis predicts a desirable reduction in traffic congestion levels in dense and developed cities such as New York City, assuming shared AVs and residents having the option to combine distinct transportation modes including rail transit, autonomous buses, autonomous shuttles, and door-to-door autonomous travel. The report suggests 25 percent of the market could be grabbed by shared AVs in 2030.[247] This compares to a predicted 30 percent share of passenger miles by private cars and privately used robotaxis which, in turn, would be less than the 35 percent for private cars in 2018.

Assuming the deployment of shared AVs—and not of private robo-taxis only—McKinsey forecasts private cars will be used less, and AV shuttles could account for a quarter of passenger-miles by 2030.[248]

McKinsey also suggests a shared AV deployment enabled by intelligent traffic systems, advanced rail signaling, and connectivity-enabled predictive maintenance would empower the AV network reliability. As such, the analysis considers all five transit system indicators—availability, affordability,

---

246 Todd Litman, *Autonomous Vehicle Implementation Predictions: Implications for Transport Planning* (Victoria, Canada: Victoria Transport Policy Institute, 2020).

247 Eric Hannon et al., "The Road to Seamless Urban Mobility," McKinsey & Company, January 16, 2019.

248 Ibid.

efficiency, convenience, and sustainability—would be improved.²⁴⁹

"It could accommodate up to 30 percent more passenger-kilometers (availability) while reducing average time per trip by 10 percent (efficiency). It could cost 25 to 35 percent less per trip (affordability), increase the number of point-to-point trips by 50 percent (convenience), and, if AVs are electric, lower GHG [greenhouse gas] emissions by up to 85 percent (sustainability)."

The coordinated traffic signals via IoT would also have a big impact, and there would be no more idling at stoplights and shorter overall trips would save electricity.

ENABLING MAAS THROUGH ELECTRIFICATION

Wow—it does not get better than that, does it? I've got news for you—of course it does! While McKinsey's 2030 envisioned transportation system includes the availability of additional transport modes, it does not assume the implementation of Mobility as a Service. Hence, McKinsey's 2030 imagined mobility landscape does not integrate all modes of available transportation types into a single digital platform. This makes me even more confident by having MaaS as a first transport system layer, the American population will see even greater transport benefits—larger reduction in traffic congestion and a more sustainable, cost-effective, and inclusive transportation system.

---

249 Ibid.

Sami Pippuri, former chief technology officer (CTO) at Global MaaS and creator of the Whim app, explained during our interview how MaaS, in fact, limits redundancy in transportation offerings and thus improves the utilization rate of all available vehicle types.

"By implementing Mobility as a Service, you can prevent that future where you have overlapping, same services being offered. When you can actually have some intellect on the top layer to see, 'okay, I don't need any more vehicles in this area, where I need them is here.' You need to have that kind of layer... That's how you can actually increase the utilization rates of those cars."

Similarly, Dominique Bonte, managing director and vice president at New York-based market research and market intelligence firm ABI Research, speaks to a higher utilization rate of vehicles as a result of a future MaaS implementation:[250]

"MaaS will result in more environmentally friendly transportation through the deployment of fleet-based, alternative powertrain vehicles and reduced congestion through improved utilization rates of available resources."

Aside from working toward seamless autonomous integrated transportation, the deployment of an electric, more sustainable mobility over the next few decades will be key. Although EVs only make a small percentage of today's overall vehicle fleet—with traditional internal-combustion

---

250 Oyster Bay, "ABI Research Forecasts Global Mobility as a Service Revenues to Exceed $1 Trillion by 2030," ABI Research, September 12, 2016.

engine (ICE) cars dominating the market—the rapid development in EV technologies and infrastructure will lead to a ramping vehicle electrification.

In another interview discussing the future of mobility, Sampo Hietanen also refers to electrification as a big phenomenon in mobility and speaks to its advantages through its implementation in shared vehicles and fleets as part of a MaaS ecosystem.[251]

"Electrification done correctly, in a way that it will be shared and electric, which we have a great chance, is actually a big empowering thing. If the electric cars will be shared and available…these [MaaS and EV] actually promote one another. So, it's not [electrification] a barrier, it's an enabler."

THE FUTURE IS NOW

We are witnessing today how companies are laying the groundwork and embracing technological advancements.

In a December 2020 online event, Toyota announced the launch of an operations management system to support services which will enable the use of an electric vehicle for autonomous Mobility as a Service. They called it the e-Palette.[252]

---

251 *TheCamp*, "Disruption in Mobility and Why it Will Be Collaborative," October 30, 2020, video, 34:38.

252 "Toyota Shows e-Palette Geared Towards Practical MaaS Application," Toyota. press release, December 22, 2020, on the Toyota website, accessed January 8, 2021.

Keiji Yamamoto, president of Toyota Connected, referred to President Akio Toyoda's announcements at the January 2018 Consumer Electronics Show (CES). Toyoda unveiled both the e-Palette and the Mobility Service Platform (MSPF) for realizing those services. Both as a symbol of mobility that goes beyond cars that provide customers services and bring new value for investors.

Yamamoto says the automated e-Palette had its debut at the 2019 Tokyo Motor Show. He claims it will provide a loop-line bus transportation service for athletes and related staff in the Olympic and Paralympic villages at the 2020 Tokyo Olympic and Paralympic Games postponed to July 2021 (stay tuned).

Yamamoto further speaks to Toyota's mission of providing people safe and comfortable mobility services and the motivation behind the deployment of the e-Palette.[253]

"We believe communities will increasingly need new mobility services such as the e-Palette and other Autono-MaaS options. The MSPF represents the framework and technology that support the provision of the e-Palette services."

"When it comes to mobility services, we assume customers expect to go where it is needed, when needed, on time, and provide the services and goods that are needed, when needed, and on time. In other words, just-in-time mobility service is required."

---

253 Ibid.

Toyota's operations management system is expected to reduce customer waiting times and alleviate congestion to ensure services provide safety, peace of mind, and comfort.

Yamamoto suggests with the objective of achieving the ultimate Toyota Production System (TPS) based just-in-time mobility service, the autonomous mobility management system (AMMS) is able to dispatch e-Palette vehicles when needed, where needed, and in the amount needed.

"Operation schedules can be changed flexibly, with vehicles automatically dispatched and returned, according to real-time mobility needs. When additional vehicles are introduced into a service, the intervals between vehicles are adjusted to ensure even spacing of services."

Yamamoto explains how, if a vehicle were to suffer damage, the system would detect it and redirect the e-Palette to the depot while ensuring operations at all times.

"Vehicle abnormalities are also automatically detected and, if that happens, the vehicles are automatically returned to the depot and replacement vehicles are immediately dispatched on the route to ensure stability of operation. In an emergency, the vehicles can be stopped and returned to service remotely, with an extra level of safety management, to provide passengers with peace of mind."

Moreover, Yamamoto affirms how, in collaboration with several partners, Toyota is also planning to deploy the e-Palettes in Woven City (Japan), a fully connected prototype city, while targeting commercial use in multiple areas and regions over the next few years.

Toyota's launch of its mobility service platform (MSPF) and e-Palette vehicle is a giant leap for the development of an electric and autonomous Mobility as a Service ecosystem. Although such initiative does not currently entail the deployment of a digital platform for the user, the idea of on-demand automated and sustainable vehicles serving communities is a huge first step. I can foresee further advancements into integrating together distinct modes of transportation to further empower seamless transportation and freedom of mobility for individuals.

Toyota's Woven City speaks to this transformation in mobility. Woven City is a human-centric, fully connected city developing technologies including autonomous driving, MaaS, robotics, and artificial intelligence in a real-world environment, and plans the deployment of e-Palette vehicles.[254,255]

"Operating within a real-world environment where people live will provide a range of lessons through which the platform will continually evolve to enable services that provide customers with safety, peace of mind, and comfort."

254 "Toyota Woven City," Toyota, accessed January 8, 2021.
255 "Toyota Shows e-Palette Geared Towards Practical MaaS Application," Toyota. press release, December 22, 2020, on the Toyota website, accessed January 8, 2021.

THE VISION FOR MOBILITY 2040

We have been praising emerging technologies like artificial intelligence, big data, and the internet of things, which has brought us to the rapid technological advancements in mobility trends including EVs, AVs, micromobility, microtransit, shared mobility, and MaaS. Now let's imagine what seamless autonomous and sustainable integrated mobility would look like for a commuter's journey in 2040:

It's 8:30 p.m. on Monday October 22nd of 2040, and after another unpredictable long day at work in downtown Washington, DC, Liliana (Lili), forty-three and a mother of two, is ready to return to her home in Alexandria (VA). As she's going down the elevator, she pulls up her smartphone and checks her best route options in the DuShare app. The app connects all of the DC metro area mobility options under one monthly subscription. Boom! Lili has got her return trip all figured out in just a few seconds—e-scooter, metro, and autonomous shuttle.

Shockingly, Lili and her family have been car-free for over a decade. Upon realizing they only used their electric SUV occasionally for weekend getaways, they got rid of it. Aside from these trips which they still enjoy through electric car rentals included in their DuShare subscription, they have relied on seamless autonomous integrated multi-modal transportation. Traffic congestion, long commutes, and competitions for a parking spot are distant memories for Lili. Parking lots and street parking have essentially vanished and have converted to bike docks and bike and scooter lanes.

For Lili's return commute, the nearest metro station is nearly a mile away from her office. While she would normally walk that distance or pull a bike from the closest docking station to get some fitness, it's late, cold, and she's tired. This time, Lili scans her phone to unlock one of the three electric scooters sitting by the building's main door. She hops on it and rides it on a dedicated bike/scooter lane for the first leg of her trip. Five minutes later she arrives to the McPherson Square station, where she jumps onto the blue line for 25 minutes to King Street-Old Town—the next closest metro station just under a mile from her house. Lili relaxes and enjoys the ride as she live streams her favorite stand-up comedy show on her phone.

As she exits and heads toward the rideshare pickup area, Lili orders an on-demand electric autonomous shuttle. She jumps in with three other passengers who happened to be on the same train, and Lili is dropped off five minutes later in front of her house. Her husband and kids welcome her with a delicious dinner.

KEY TAKEAWAYS

- Electrification and automation are the ultimate enablers of a Mobility as a Service (MaaS) transport ecosystem.
- The future of transportation in the United States is seamless, autonomous, and sustainable multi-modal integrated mobility.
- While such a transport system will not be deployed overnight, we are already witnessing today, through the case of Toyota's electric and autonomous car for MaaS, the

laying of groundwork toward a successful amalgamation and integration of mobility trends.

- Although the above envisioning of Mobility 2040 speaks to a commute between urban and suburban areas, I am a strong believer the US transportation system can and will evolve to provide these benefits to the entire population. The transport system will minimize traffic congestion, long commutes, vehicular crashes, air pollution, and especially bridge the transportation accessibility gap.

CONCLUSION

"Transportation is the center of the world. It is the glue of our daily lives."[256]

—ROBIN CHASE, COFOUNDER OF ZIPCAR

The modern history of the United States transportation system has taught us that in light of its rapid development through the creation of the Interstate Highway System in the 1950s, the subsequent suburban expansion lacked direction and focus. This has translated into a high vehicle dependency, decreased levels of public transit ridership, and an overall lack of investment in public transport.

As a result, the US transportation system faces a myriad of pressing challenges including traffic congestion, long commutes, vehicular crashes, parking difficulties, high levels of vehicular carbon emissions, and transportation accessibility.

---

256 Robin Chase, quoted in "Reimagining Public Transit", Mission, *Medium*, December 14, 2018.

Fortunately, emerging technologies like the internet of things (IoT), 5G connectivity, cloud computing, big data, and artificial intelligence (AI) are providing the infrastructure that will enable larger scale improvements in the transportation sector. Data coming from people's mobile devices is being leveraged for transportation planning, among other use cases. Innovative companies and startups are playing a key role in disrupting the transportation sector, and distinct mobility trends are surging from the intersection of cutting-edge technologies and transportation.

Micromobility has emerged as a disruptive transport alternative offering more sustainable modes of transportation to private cars. Due to its relative affordability and convenience, micromobility options—including bikesharing systems and electric scooters—are increasingly being perceived as an effective first- and last-mile transportation method. Public-private partnerships are being established toward the deployment of micromobility options. Cities are further promoting the implementation of infrastructure (e.g., bike lanes) necessary for the safe deployment of micromobility solutions. Micromobility service providers, cities and jurisdictions are increasing their efforts to make rideables accessible to all residents.

Leveraging data and technology, microtransit appears as a cost-effective, sustainable, convenient, and inclusive transportation method for rural, suburban, and urban US communities. Microtransit vehicles are especially effective in suburban and rural areas when they are used as additions to existing underfunded transit systems. They work well serving first- and last-mile connections, and enabling people access to jobs, resources, and other social privileges. In urban areas, microtransit is helping reduce personal car driven miles, associated traffic congestion, and vehicular carbon emissions. Microtransit initiatives are also becoming more popular through public-private partnerships, involving the collaboration of a variety of stakeholders including transit agencies, and mobility and technology companies.

Shared mobility has appeared as an innovative mobility solution to rapid US urbanism, associated increased traffic congestion, and environmental pollution. While the US transportation system was built around the privately owned car, the everlasting automobile dependency is progressively shifting toward new available on-demand shared transportation modes.

Mobility as a Service (MaaS) is the future, and the solution to the rapid urbanization in the US. It consists of the movement from personally owned vehicles toward mobility solutions provided by a combination of public and private organizations. MaaS is a digital platform where you can plan, book, and pay for a travel journey all at once. By offering a myriad of transportation options, MaaS is a great alternative to the personal car for those who don't own a car, can't drive, or simply would like to leave their vehicles at home. More so, MaaS is about ensuring freedom of mobility by offering choice.

MaaS comes off as cost-effective transportation, especially when compared to the elevated car ownership costs, and holds the promise of being a more sustainable transport system alternative to the car. Over the past few years, several private companies/startups and public transit entities have been exploring and investing in the development of MaaS solutions. While most of these MaaS services have an urban/ city focus, MaaS solutions can also be deployed in suburban and rural areas and at a regional level. Even though MaaS has been popularized mainly in Europe, pilot programs on integrated on-demand mobility have also been developing in the United States.

Given the early stage of MaaS, it will need, however, to undergo a period of development prior to reaching maturity and full implementation. Key areas need to be fully explored and the right questions asked and answered on the path to a definite MaaS solution deployment in the United States. Exploratory areas of interest include further carving out a space for MaaS in the transport ecosystem, building effective partnerships across the private/public sectors, managing the critical balance between innovation and regulation, and bridging the US transportation infrastructure gap.

Despite the technological, governance, and regulatory barriers MaaS foresees, there are numerous organizations and private-public sector partnerships promoting and advancing the vision, agenda, and development of MaaS in the US and the world. Examples include MaaS Alliance in Europe, MaaS America in the US, TRB, ASCE, and APTA, among others.

On the path to the future US transportation system—seamless autonomous and sustainable multi-modal integrated mobility—electrification and automation appear as the ultimate enablers of a MaaS transport ecosystem. While such a transport system will not be deployed overnight, we are already witnessing key stakeholders laying the groundwork toward a successful amalgamation and integration of mobility trends.

I believe we can reinvent US transportation systems over the next two decades so it benefits the entire population. We can do this by leveraging cutting-edge technologies and motivated stakeholders already in place. This, however, will be hardly possible without the collaboration of you, the readers of this book, whether you are investors, policy makers, transportation officials, or simply commuters who can vote with your purchasing power. This shift will come down to all of us.

Mobility 2040 is a future that belongs to us all, and the creation of that future starts now. The first step is recognizing and accepting the present transport-related problems, along with private and public sectors sharing the mission of transforming the US transportation system while responding to these environmental and societal concerns. I am convinced with all stakeholders working collectively and toward the same direction, Mobility 2040 will become fully realized in the US.

To private transportation service and mobility companies, original equipment manufacturers, consumers (commuters), public transit agencies, regulators, policy makers, and investors:

Let's create a future where Lili can decide to leave work on a whim, change her mind on preferred mode of transportation, live stream her favorite comedy show on the train ride home, and finish her last mile commute in a low-cost, on-demand shuttle service right to her front door.

ACKNOWLEDGMENTS

In creating this book, I had the wonderful opportunity to interview many bright and inspiring men and women within the transportation industry. They all shared with me their unique insights, opinions, and expertise. Each conversation taught me something new, and each person brought a different perspective and presented a new way of thinking about mobility.

To all my interviewees, thank you for making this project possible. Your passion and commitment have bolstered my curiosity to keep asking questions.

Special thanks go to my terrific beta readers who provided me invaluable feedback during the revision process. Your comments and suggestions were priceless. I am grateful for you becoming a part of my intense and yet extremely rewarding journey.

I want to thank each and every person who pre-ordered the e-book, paperback, and multiple copies to make publishing possible. Thank you to everyone who helped spread the word about *Mobility 2040* to create amazing momentum. I was overwhelmed by the amount of support and encouragement I received during this fascinating process.

My eternal gratitude to the following readers:

Abhinit Pandey
Akash Kang
Alain Descamps
Alexandra Boycheck
Ali Carpenter
Amy Setton
Andrea Armijos
Andreea Russo
Andrei Nitica
Andres Romero
Andres S. Esquetini
Andrew Gordon
Andy Fernandez
Anjali Shahani
Annia Bowen
Audrey Del Rosario
Austin McCarty
Benjamin King
Berbyn Levy
Bianca Lopez de Victoria
Blerina Krasniq
Bryan Crosson
Camelia Sporea
Camila Torres
Cesar Cantos
Charlie Kao

Chong Hwan Kim
Christina Carpenter
Christine Nieva
Claudia Ratti
Clotilde Bowen*
Cristina Ratti
Damian Saccocio
Daniel Capriles
Daniela Rodriguez
Dara Novini
Dustin Gallegos
Eduardo Tirado
Elias Rivera
Emilio Bowen
Emily Ferrell
Enrico Dinges
Eric Koester
Eric Messinger
Estevan Astorga
Fatima Tellez
Fernando Armendaris
Francisco Flores
Francisco Perez
Francisco Serna
Frank Sariol
Fredy Bustamante

Gabriel Veras
Gabriela Miranda
Galia Bustamante*
Galo Bustamante
Galo F. Bustamante
Gerard Aretakis
Giovanna De Maio
Giulliana Ratti
Gordon Bradshaw
Gregory Cole
Guillermo Dahik
Homero Soto
Humberto Heredia
Iolani L. Bullock
Ivan Ruilova
Jared Koch
Jason Chang
Javier Bowen
Jeff Reilly
Jeffrey DeBoer
Jennifer Bohlander
Jennifer Vargas
Jesse Castro
John Leatherman
Jordan Hull*
Jordan Jarrett
Jose Andrade
Juan Cadena
Juan Pablo Bastidas
Juan Pablo Celis
Julia Rosenthal
Julio C. Monroy Cruz*

Junior Mwemba
Kaitlyn Smith
Karina Sicherle*
Katherine Vargas
Keisuke Adachi
Kelly Ambrose
Kelsey Freeman
Kelsey Knapp
Lawrence J. Verbiest
Louie Borja
Lucia Valverde
Luly Menoscal
Lydia Kickham
Marc Punnette
Marco Guevara
Maria Isabel Bustamante*
Maria Mercedes Blanco
Maria Palacio
Maria Pilar Zambrano
Mariana Fernandez
Mariana Pestana
Mariecarmen Aza
Mark Vogel
Michelle Wu
Miguel Saavedra
MJ Bustamante*
Mohit Gupta
Monica Paladines
Natalie Gonzalez
Negar Ahsan
Newton Davis
Olena Prykhodko

Oren Katz
Pablo Cabrera
Pamela Ponce
Paola Jurado
Parizad Edulbehram
Patricia Bowen
Patricia Rivera
Prashant Malaviya*
Rafael Heredia Lucio
Rafic El Helou
Raluca Chelaru
Raluca David
Rebecca Norman
Reynaldo Ratti
Ricardo Rendon Cepeda
Riccardo Moauro
Rishikesh Moharkar
Robert Aretakis
Robert Fernandez
Rochelle Fin
Rodion Stolyar
Rohan Dalvi
Rohit Lala
Ronald W. Rye

Rowland Henshaw
Ryan and Erin Fehr
Saleh Aldrees
Samuel Kim
Sandra Grega
Sania Mohammed
Santi Blando
Santiago Mayoral
Sara Eisenberg
Saskia van de Bilt
Sebastian Mayor
Shainur Ahsan
Stefanie Cohen
Stephanie Burns
Stephanie Gonzalez
Teresa Lamar*
Thais Ferreira
Timothy Ballenger
Tomas Facio Pacheco
Tomas Valdes
Ugne Buivydaite
Utsab Khadka
Varun Nagarajan
Vicky Fernandez

*multiple copies/campaign contributions

Finally, I would like to acknowledge New Degree Press and my incredible publishing and support team: Eric Koester, Ryan Porter, Judy Rosen, Brian Bies, Venus Bradley, Leila Summers, Amanda Brown, Elina Oliferovskiy, Mateusz Cichosz and Gjorgji Pejkovski.

APPENDIX

INTRODUCTION

American Public Transportation Association (APTA). "Public Transportation Facts." Accessed September 12, 2020. https://www.apta.com/news-publications/public-transportation-facts/.

Center for Automotive Research (CAR). "Disrupted by Mobility Startups, Automakers Reshape Their Roles." May 4, 2018. https://www.cargroup.org/disrupted-by-mobility-startups-automakers-reshape-their-roles/.

Hartikainen, Ali, Jukka-Pekka Pitkänen, Atte Riihelä, Jukka Räsänen, Ian Sacs, Ari Sirkiä, and Andre Uteng. "Whimpact." *Ramboll.* May 21, 2019. https://ramboll.com/-/media/files/rfi/publications/Ramboll_whimpact-2019.pdf.

Mibiz. "Pilot Program to Boost Mobility Options for People with Disabilities." July 13, 2020. https://mibiz.com/sections/economic-development/pilot-program-to-boost-mobility-options-for-people-with-disabilities.

Slone, Sean. "Top 5 Issues for 2018: Transportation & Infrastructure: The Precarious Condition of US Infrastructure." *Sean Slone's blog*. *The Council of State Governments*, January 21, 2018. https://knowledgecenter.csg.org/kc/content/top-5-issues-2018-transportation-infrastructure-precarious-condition-us-infrastructure.

The Rapid. "New App Offers The Rapids GO!Bus Passengers Convenience and Less Wait Time." The Rapid. press release, August 5, 2019. The Rapid website. Accessed September 8, 2020. https://www.ridetherapid.org/articles/new-app-offers-the-rapids-gobus-passengers-convenience-and-less-wait-time.

Uber Technologies, Inc. "Uber Announces Results for Fourth Quarter and Full Year 2020." Uber Investor. Uber Technologies, Inc. press release, February 10, 2021. Uber Technologies, Inc. website. Accessed February 12, 2021. https://investor.uber.com/news-events/news/press-release-details/2021/Uber-Announces-Results-for-Fourth-Quarter-and-Full-Year-2020/default.aspx.

United States Environmental Protection Agency (EPA). "Sources of Greenhouse Emissions." Last modified December 4, 2020. https://www.epa.gov/ghgemissions/sources-greenhouse-gas-emissions.

United States Environmental Protection Agency (EPA). "Transportation and Climate Change." Carbon Pollution from Transportation. Last modified November 20, 2020. https://www.epa.gov/transportation-air-pollution-and-climate-change/carbon-pollution-transportation.

CHAPTER 1: THE MODERN HISTORY OF THE US TRANSPORTATION SYSTEM

Benioff, Marc. Quoted in Scott D. Hariss. "2009 Q&A: Marc Benioff, CEO of Salesforce.com." *The Mercury News.* October 23, 2009. https://www.mercurynews.com/2009/10/23/2009-qa-marc-benioff-ceo-of-salesforce-com/.

Buehler, Ralph. "9 Reasons the US Ended Up So Much More Car-Dependent Than Europe." Bloomberg CityLab. *Bloomberg.* February 4, 2014. https://www.bloomberg.com/news/articles/2014-02-04/9-reasons-the-u-s-ended-up-so-much-more-car-dependent-than-europe.

Burwell, David, and Shin-Pei Tsay. "Transforming Transportation for the 21st Century." Carnegie Endowment for International Peace. July 14, 2011. https://carnegieendowment.org/2011/07/14/transforming-transportation-for-21st-century.

Florida, Richard. "The Best and Worst US Places to Live Car-Free." Bloomberg CityLab. *Bloomberg.* September 24, 2019. https://www.bloomberg.com/news/articles/2019-09-24/the-best-and-worst-u-s-places-to-live-car-free.

Jaffe, Eric. "These 2 Charts Prove American Drivers Don't Pay Enough for Roads." Bloomberg CityLab. *Bloomberg.* September 18, 2013. https://www.bloomberg.com/news/articles/2013-09-18/these-2-charts-prove-american-drivers-don-t-pay-enough-for-roads.

NationMaster. "Motor Vehicles Per 1000 People: Countries Compared." Last modified 2014. https://www.nationmaster.com/

country-info/stats/Transport/Road/Motor-vehicles-per-1000-people.

Norton, Peter. Quoted in Joseph Stromberg. "The Real Story Behind The Demise of Americas Once-Mighty Streetcars." *Vox.* May 7, 2015. https://www.vox.com/2015/5/7/8562007/streetcar-history-demise.

Statista. "Gasoline Prices In Selected Countries Worldwide In 4th Quarter of 2019." Accessed September 10, 2020. https://www.statista.com/statistics/221368/gas-prices-around-the-world/.

Stromberg, Joseph. "The Real Reason American Public Transportation Is Such A Disaster." *Vox.* Last modified August 10, 2015. https://www.vox.com/2015/8/10/9118199/public-transportation-subway-buses.

United States Department of Transportation Federal Highway Administration. "Highway Statistics 2017." Policy and Governmental Affairs Office of Highway Policy Information. Last modified November 27, 2018. https://www.fhwa.dot.gov/policyinformation/statistics/2017/hm20.cfm.

United States Department of Transportation Federal Highway Administration. "State Motor-Vehicle Registrations 2018." Policy and Governmental Affairs Office of Highway Policy Information. December, 2019. https://www.fhwa.dot.gov/policyinformation/statistics/2018/pdf/mv1.pdf.

CHAPTER 2: PROBLEMS WITH THE SYSTEM TODAY

American Automobile Association (AAA). "Your Driving Costs, 2020." December 14, 2020. https://newsroom.aaa.com/wp-content/uploads/2020/12/Your-Driving-Costs-2020-Fact-Sheet-FINAL-12-9-20-2.pdf.

American Lung Association. "More than 4 in 10 Americans Live with Unhealthy Air; Eight Cities Suffered Most Polluted Air Ever Recorded." American Lung Association. press release, April 24, 2019. American Lung Association website. Accessed August 25, 2020. https://www.lung.org/media/press-releases/sota-2019.

Blincoe, Lawrence J., Miller, Ted R., Zaloshnja, Eduard, and Lawrence, Bruce A. *The Economic and Societal Impact of Motor Vehicle Crashes, 2010.* Report No. DOT HS 812 013. Washington, DC: National Highway Traffic Safety Administration, 2015. https://crashstats.nhtsa.dot.gov/Api/Public/ViewPublication/812013.

IQAir. "World's most polluted countries 2019 (PM2.5)." Accessed September 25, 2020. https://www.iqair.com/us/world-most-polluted-countries.

Jiao, Junfeng and Chris Bischak. "Dozens of US Cities Have Transit Deserts Where People Get Stranded." *Smithsonian Magazine.* March 16, 2018. https://www.smithsonianmag.com/innovation/dozens-us-cities-have-transit-deserts-where-people-get-stranded-180968463/.

Long, Heather. "A Record 7 Million Americans Are 3 Months Behind On Their Car Payments, A Red Flag For The Economy."

The Washington Post. February 12, 2019. https://www.washingtonpost.com/business/2019/02/12/record-million-americans-are-months-behind-their-car-payments-red-flag-economy/.

Loveday, Steven. "Electric Cars With the Longest Range in 2021." US News & World Report. September 16, 2020. https://cars.usnews.com/cars-trucks/electric-cars-with-the-longest-range.

Miller Kory Rowe LLP. "Motor Vehicle Crashes Cost the US Nearly $1 Trillion/Year." April 27, 2017. https://www.mkrfirm.com/blog/motorcycle-accidents/motor-vehicle-crashes-cost-u-s-nearly-1-trillionyear/.

National Academies of Sciences, Engineering, and Medicine. *Critical Issues in Transportation 2019.* Washington, DC: The National Academies Press, 2018. https://doi.org/10.17226/25314.

National Center for Statistics and Analysis. *Overview of Motor Vehicle Crashes in 2019.* Traffic Safety Facts Research Notes. Report No. DOT HS 813 060. Washington, DC: National Highway Traffic Safety Administration, 2020. https://crashstats.nhtsa.dot.gov/Api/Public/ViewPublication/812826.

Peñalosa, Enrique. Quoted in Matthew Roth. "Enrique Peñalosa Urges SF to Embrace Pedestrians and Public Space." *StreetsBlog SF.* July 8, 2009. https://sf.streetsblog.org/2009/07/08/enrique-penalosa-urges-sf-to-embrace-pedestrians-and-public-space/.

Reed, Trevor. "INRIX Global Traffic Scorecard." INRIX. INRIX Research. March, 2020. https://www.scribd.com/document/473840962/INRIX-2019-Traffic-Scorecard-WEB.

TEDx Talks. "The Air We Breathe Is Killing Us—But It Doesn't Have To | Beth Gardiner | TEDxLondon." June 11, 2019. Video, 12:28. https://www.youtube.com/watch?v=gXDAGYLu3X8.

Union of Concerned Scientists. "Cars, Trucks, Buses and Air Pollution." Last modified July 19, 2018. https://www.ucsusa.org/resources/cars-trucks-buses-and-air-pollution.

Union of Concerned Scientists. "Each Country's Share of CO2 Emissions." Last modified August 12, 2020. https://www.ucsusa.org/resources/each-countrys-share-co2-emissions.

United Nations. "68% of the world population projected to live in urban areas by 2050, says UN." May 16, 2018. https://www.un.org/development/desa/en/news/population/2018-revision-of-world-urbanization-prospects.html#.

United States Environmental Protection Agency (EPA). "AQI Air Quality Index." Office of Air Quality Planning and Standards. EPA-456/F-14-002. February, 2014. https://www.airnow.gov/sites/default/files/2018-04/aqi_brochure_02_14_0.pdf.

United States Environmental Protection Agency (EPA). "Sources of Greenhouse Emissions." Last modified December 4, 2020. https://www.epa.gov/ghgemissions/sources-greenhouse-gas-emissions.

Williams, Joseph P. "In an Unequal America, Getting to Work Can Be Hell." *The Nation.* January 29, 2019. https://www.thenation.com/article/archive/transit-deserts-extreme-commuters-inequality/.

CHAPTER 3: ELECTRIC VEHICLES

Baldwin, Roberto. "Tesla Is Working on 620-Mile Range for Future Cars, Upcoming Semi." *Car and Driver.* November 24, 2020. https://www.caranddriver.com/news/a34775237/elon-musk-tesla-car-semi-range/#sidepanel.

Center for Climate and Energy Solutions (C2ES). "Reducing Your Transportation Footprint." Accessed October 5, 2020. https://www.c2es.org/content/reducing-your-transportation-footprint/.

Downing, Shane. "8 electric truck and van companies to watch in 2020." GreenBiz. January 13, 2020. https://www.greenbiz.com/article/8-electric-truck-and-van-companies-watch-2020.

DPCcars. "Elon Musk Explains Tesla Motors Electric Vehicle History." June 12, 2016. Video, 19:47. https://www.youtube.com/watch?v=QMpWXN8IC-U.

General Motors. "General Motors Future Electric Vehicles to Debut Industry's First Wireless Battery Management System." News. September 9, 2020. https://media.gm.com/media/us/en/gm/news.detail.html/content/Pages/news/us/en/2020/sep/0909-wbms.html.

General Motors. "GM Reveals New Ultium Batteries and a Flexible Global Platform to Rapidly Grow its EV Portfolio." News. March 4, 2020. https://media.gm.com/media/us/en/gm/home.detail.html/content/Pages/news/us/en/2020/mar/0304-ev.html.

JuiceBlog (blog). "Understanding the Different EV Charging Levels." Enel X. May 8, 2019. https://evcharging.enelx.com/resources/blog

Loveday, Steven. "Electric Cars With the Longest Range in 2021." US News & World Report. September 16, 2020. https://cars.usnews.com/cars-trucks/electric-cars-with-the-longest-range.

McKerracher, Colin, Ali Izadi-Najafabadi, Aleksandra ODonovan, Nick Albanese, Nikolas Soulopolous, David Doherty, Milo Boers et al. "Electric Vehicle Outlook 2020." BloombergNEF. *Bloomberg.* 2020. https://about.bnef.com/electric-vehicle-outlook/.

MyBroadband. "Tesla is Now More Valuable Than 12 Top Automakers Combined." November 19, 2020. https://mybroadband.co.za/news/motoring/376531-tesla-is-now-more-valuable-than-12-top-automakers-combined.html.

Musk, Elon. Quoted in Lorraine Chow. "Elon Musk: You Can Easily Power All of China With Solar." *EcoWatch.* January 29, 2016. https://www.ecowatch.com/elon-musk-you-can-easily-power-all-of-china-with-solar-1882162228.html#toggle-gdpr.

Palmer, Annie. "Amazon Debuts Electric Delivery Vans Created with Rivian." *CNBC.* Last modified October 8, 2020. https://www.cnbc.com/2020/10/08/amazon-new-electric-delivery-vans-created-with-rivian-unveiled.html.

Plungis, Jeff. "How the Electric Car Charging Network Is Expanding." Consumer Reports. Last modified November 12, 2019.

https://www.consumerreports.org/hybrids-evs/electric-car-charging-network-is-expanding/.

Popple, Ryan. "#055: Electric Buses 101 with Ryan Popple, CEO, Proterra." October 18, 2019. In *Mobility Podcast*. Produced by Greg Rogers. Podcast, Soundcloud, 40:10. https://soundcloud.com/user-223028423/055-electric-buses-101-with-ryan-popple-ceo-proterra.

StreetLight Data. "EV Infrastructure Planning." Accessed August 9, 2020. https://www.streetlightdata.com/electric-vehicle-charging-station-deployment-data.

The White House. "Fact Sheet: Obama Administration Announces Federal and Private Sector Actions to Accelerate Electric Vehicle Adoption in the United States," The White House. press release, July 21, 2016. The White House website. Accessed January 25, 2021. https://obamawhitehouse.archives.gov/the-press-office/2016/07/21/fact-sheet-obama-administration-announces-federal-and-private-sector.

Union of Concerned Scientists. "Cars, Trucks, Buses and Air Pollution." Last modified July 19, 2018. https://www.ucsusa.org/resources/cars-trucks-buses-and-air-pollution.

United States Department of Energy. "Electric Vehicle Basics." Office of Energy Efficiency & Renewable Energy. Accessed October 5, 2020. https://www.energy.gov/eere/electricvehicles/electric-vehicle-basics.

United States Department of Energy. "President Obama Announces $2.4 Billion in Funding to Support Next Generation Electric

Vehicles." March, 2019. https://www.energy.gov/articles/president-obama-announces-24-billion-funding-support-next-generation-electric-vehicles.

United States Environmental Protection Agency (EPA). "Sources of Greenhouse Emissions." Last modified December 4, 2020. https://www.epa.gov/ghgemissions/sources-greenhouse-gas-emissions.

USA Facts. "How Many Electric Cars Are On The Road In The United States?" Last modified October 22, 2020. https://usafacts.org/articles/how-many-electric-cars-in-united-states/#.

Vincent, John M. "How Does the Electric Car Tax Credit Work?" US News & World Report. June 15, 2020. https://cars.usnews.com/cars-trucks/how-does-the-electric-car-tax-credit-work.

Virta. "Smart Charging of Electric Vehicles." Accessed January 11, 2021. https://www.virta.global/smart-charging#.

Wagner, I. "Electric vehicle charging stations and outlets in US - February 2021." Statista. February 16, 2021. https://www.statista.com/statistics/416750/number-of-electric-vehicle-charging-stations-outlets-united-states/.

CHAPTER 4: EMERGING TECHNOLOGIES FOR A BRIGHTER FUTURE

Aeris. "How IoT Is Changing Sleep Therapy." *Blog.* Accessed September 8, 2020. https://www.aeris.com/news/post/how-iot-is-changing-sleep-therapy/.

AT&T. "How 5G Will Impact the Transportation Industry." AT&T Business. Accessed September 9, 2020. https://www.business. att.com/learn/tech-advice/how-5g-will-impact-the-transportation-industry.html.

Bezos, Jeff. Quoted in Express Information Systems. "Four Key Trends Reshaping Wealth & Asset Management Accounting." November 9, 2020. https://www.expressinfo.com/four-key-trends-reshaping-wealth-asset-management-accounting/.

BrightTALK. "Founders Spotlight—Episode 6: Laura Schewel, Founder and CEO, Streetlight Data." May 9, 2019. Video: 44:35. https://www.youtube.com/watch?v=aqnY5jOMeoI.

Burgess, Matt. "What is the Internet of Things? WIRED Explains." *WIRED.* February 16, 2018. https://www.wired.co.uk/article/internet-of-things-what-is-explained-iot.

Grand View Research. "Global 5G Services Market Size Report, 2021–2027." May, 2020. https://www.grandviewresearch.com/industry-analysis/5g-services-market#.

Investopedia. s.v. "Artificial Intelligence (AI)." By Jake Frankenfield. Accessed September 9, 2020. https://www.investopedia.com/terms/a/artificial-intelligence-ai.asp.

Joshi, Naveen. "How AI Can Transform The Transportation Industry." *Forbes.* July 26, 2019. https://www.forbes.com/sites/cognitiveworld/2019/07/26/how-ai-can-transform-the-transportation-industry/.

McGarry, Caitlin. "5G vs 4G Performance Compared." Tom's Guide. February 25, 2021. https://www.tomsguide.com/features/5g-vs-4g.

Mordor Intelligence. "IoT in Transportation Market—Growth, Trends, Covid-19 Impact, and Forecasts (2021 - 2026)." Accessed September 9, 2020. https://www.mordorintelligence.com/industry-reports/iot-in-transportation-market.

Oracle. "Big Data Defined." Accessed September 9, 2020. https://www.oracle.com/big-data/what-is-big-data/.

Overby, Stephanie. "How to Explain Edge Computing in Plain English." The Enterprisers Project. November 30, 2020. https://enterprisersproject.com/article/2019/7/edge-computing-explained-plain-english.

Reed, Trevor. "INRIX Global Traffic Scorecard." INRIX. INRIX Research. March, 2020. https://www.scribd.com/document/473840962/INRIX-2019-Traffic-Scorecard-WEB.

TEDx Talks. "Drive for Change | Laura Schewel | TEDxSacramentoSalon." August 11, 2015. Video, 11:00. https://www.youtube.com/watch?v=C8hF8y-XpVY.

Simplilearn. "Artificial Intelligence & the Future—Rise of AI (Elon Musk, Bill Gates, Sundar Pichai)|Simplilearn." March 26, 2019. Video, 4:51. https://www.youtube.com/watch?v=wTbrkosuwbg.

Synarion Solutions. "Impact Of AI, IoT And Big Data On Transportation Industry." *Blog.* Accessed September 9, 2020. https://

www.synarionit.com/blog/impact-of-ai-iot-and-big-data-on-transportation-industry/.

TechHQ. "Transportation Takes A Leading Edge with Smart Technology." July 21, 2020. https://techhq.com/2020/07/transportation-takes-a-leading-edge-with-smart-technology/#9.

Verizon. "What Does 5G mean?" Personal Tech. November 5, 2019. https://www.verizon.com/about/our-company/5g/what-does-5g-mean.

CHAPTER 5: AUTONOMOUS VEHICLES

Allied Market Research. "Autonomous Vehicle Market Outlook—2026." Accessed January 15, 2020. https://www.alliedmarketresearch.com/request-sample/4649.

Association for Safe International Road Travel (ASIRT). "Road Safety Facts." Accessed January 4, 2021. https://www.asirt.org/safe-travel/road-safety-facts/.

Ayers, Whitlow & Dressler (blog). "NHTSA: Nearly All Car Crashes Are Due To Human Error." Accessed September 10, 2020. https://www.ayersandwhitlow.com/blog/2018/01/nhtsa-nearly-all-car-crashes-are-due-to-human-error/.

Bejarano, Paula A. "Robo-Trucks: Could they be first?" *Medium.* July 21, 2019. https://medium.com/autonomousity-autonomous-vehicles-business-models/robo-trucks-could-they-be-first-3d224218b382.

Brumbaugh, Stephen. "Travel Patterns of American Adults with Disabilities." US Department of Transportation. Last modified December 11, 2018. https://www.bts.gov/travel-patterns-with-disabilities#.

Centers for Disease Control and Prevention (CDC). "Cost Data and Prevention Policies." National Center for Injury Prevention and Control. Last modified November 2, 2020. https://www.cdc.gov/transportationsafety/costs/index.html.

Estrada, David. "#065: David Estrada, Nuro." February 18, 2020. In *Mobility Podcast*. Produced by Greg Rogers. Podcast, Apple Podcasts, 1:01:13. https://podcasts.apple.com/us/podcast/065-david-estrada-nuro/id1301517009?i=1000466022847.

Etherington, Darrell. "Over 1,400 Self-Driving Vehicles Are Now In Testing By 80+ Companies Across The US." TechCrunch, June 11, 2019. https://techcrunch.com/2019/06/11/over-1400-self-driving-vehicles-are-now-in-testing-by-80-companies-across-the-u-s/.

Fairfax County. "Fairfax County and Dominion Energy Launch Public Service on Virginias First Publicly Funded Autonomous Electric Shuttle Pilot Project." October 22, 2020. https://www.fairfaxcounty.gov/transportation/news/t21_20.

GlobeNewswire. "Healthcare CRM Market Size to Exceed USD 21.46 Billion by 2025, at 13.4% CAGR, Says Market Research Future (MRFR)." Market Research Future. February 8, 2021. https://www.globenewswire.com/news-release/2021/02/08/2171259/0/en/Healthcare-CRM-Market-Size-to-Exceed-USD-21-46-Billion-by-2025-at-13-4-CAGR-Says-Market-Research-Future-MRFR.html.

GreyB. "Top 30 Self Driving Technology and Car Companies." Accessed October 23, 2020. https://www.greyb.com/autonomous-vehicle-companies/.

Hall, Chris. "Self-Driving Cars: Autonomous Driving Levels Explained." Pocket-lint. August 19, 2020. https://www.pocket-lint.com/cars/news/143955-sae-autonomous-driving-levels-explained.

Kambria. "The History and Evolution of Self-Driving Cars." June 23, 2019. https://kambria.io/blog/the-history-and-evolution-of-self-driving-cars/.

Kerry, Cameron F. and Jack Karsten. "Gauging Investment In Self-Driving Cars." The Brookings Institution. October 16, 2017. https://www.brookings.edu/research/gauging-investment-in-self-driving-cars/.

Musson, Melanie. "Which states allow self-driving cars?" Auto Insurance. Last modified February 26, 2021. https://www.autoinsurance.org/which-states-allow-automated-vehicles-to-drive-on-the-road/.

National Center for Statistics and Analysis. *Overview of Motor Vehicle Crashes in 2019.* Traffic Safety Facts Research Notes. Report No. DOT HS 813 060. Washington, DC: National Highway Traffic Safety Administration, 2020. https://crashstats.nhtsa.dot.gov/Api/Public/ViewPublication/812826.

National Highway Traffic Safety Administration. "Automated Vehicles for Safety." Accessed October 10, 2020. https://www.nhtsa.gov/technology-innovation/automated-vehicles.

OKane, Sean. "Zoox Unveils A Self-Driving Car That Could Become Amazons First Robotaxi." *The Verge*. December 14, 2020. https://www.theverge.com/2020/12/14/22173971/zoox-amazon-robotaxi-self-driving-autonomous-vehicle-ride-hailing.

Oremus, Will. "Tesla May Build Its Own Self-Driving Cars, But Prefers the Term Autopilot." *Slate*. May 7, 2013. https://slate.com/technology/2013/05/tesla-self-driving-cars-ceo-elon-musk-prefers-camera-based-autopilot-system.html#.

Slovick, Murray. "TuSimple Completes Self-Driving Truck Test for the USPS." Electronic Design. June 24, 2019. https://www.electronicdesign.com/markets/automotive/article/21808184/tusimple-completes-selfdriving-truck-test-for-the-usps.

State of California Department of Motor Vehicles. "AV Permit Holders Report Nearly 2.9 Million Test Miles in California." Office of Public Affairs. February 26, 2020. https://www.dmv.ca.gov/portal/news-and-media/av-permit-holders-report-nearly-2-9-million-test-miles-in-california/.

Ward, Tom. "Walmart and Gatik Go Driverless in Arkansas and Expand Self-Driving Car Pilot to a Second Location." Walmart. December 15, 2020. https://corporate.walmart.com/newsroom/2020/12/15/walmart-and-gatik-go-driverless-in-arkansas-and-expand-self-driving-car-pilot-to-a-second-location#.

Wiggers, Kyle. "Autonomous Vehicles Should Benefit Those with Disabilities, but Progress Remains Slow." VentureBeat. August 21, 2020. https://venturebeat.com/2020/08/21/autonomous-vehicles-disabilities-accessibility-inclusive-design/.

CHAPTER 6: MICROMOBILITY

Capital Bikeshare. "East of the Anacostia River Network Expansion." Accessed September 25, 2020. https://www.capitalbikeshare. com/blog/east-of-the-anacostia-river-network-expansion.

Capital Bikeshare. "Single Trip." Accessed September 23, 2020. https://www.capitalbikeshare.com/pricing/single-trip.

City of Portland, Oregon. "PBOT Announces New Biketown Agreement With Lyft And An Extension of Its Title Sponsorship With Founding Partner Nike, Inc. for Portland Bike-Share Through 2025." July 16, 2020. https://www.portland.gov/transportation/news/2020/7/16/pbot-announces-new-biketown-agreement-lyft-and-extension-its-title.

City of Seattle. *New Mobility Playbook*. Seattle, WA: Seattle Department of Transportation (SDOT), 2017. https://www.seattle.gov/ Documents/Departments/SDOT/NewMobilityProgram/New-Mobility_Playbook_9.2017.pdf.

Dickey, Megan Rose. "The Electric Scooter Wars of 2018." TechCrunch. December 23, 2018. https://techcrunch.com/2018/12/23/ the-electric-scooter-wars-of-2018/.

Fox, Laura. "82: The Biggest Bikeshare In America—Talking with Laura Fox, Lyfts General Manager for Citi Bike in New York."

July 23, 2020. In *Micromobility Podcast*. Produced by Oliver Bruce. Podcast, 1:07:03.

Government. *DDOT Releases New Permit Application for Dockless Vehicles*. Washington, DC: District Department of Transportation (DDOT), 2018. Quoted in Lewis, Rebecca and Rebecca Steckler. *Emerging Technologies and Cities: Assessing the Impacts of New Mobility on Cities*. Report No. NITC-RR-1249. Portland, OR: Transportation Research and Education Center (TREC), 2020. https://www.portlandoregon.gov/transportation/article/709719.

Hammock, Rex. "Einstein Explains Why Life is Like Riding a Bicycle." Motivation. SmallBusiness. November 4, 2016. https://smallbusiness.com/monday-morning-motivation/einstein-quotation-bicycle/.

Heineke, Kersten, Benedikt Kloss, Darius Scurtu, and Florian Weig. "Micromobilitys 15,000-Mile Checkup." McKinsey & Company. January 29, 2019. https://www.mckinsey.com/industries/automotive-and-assembly/our-insights/micromobilitys-15000-mile-checkup#.

Institute for Transportation & Development Policy (ITDP). "Defining Micromobility." Multimedia. Accessed September 22, 2020. https://www.itdp.org/multimedia/defining-micromobility/.

Lahoti, Nitin. "Micromobility: The Next Wave of Eco-Friendly Transportation." *Blog*. Mobisoft Infotech. October 14, 2019. https://mobisoftinfotech.com/resources/blog/future-of-micromobility/.

Lewis, Rebecca and Rebecca Steckler. *Emerging Technologies and Cities: Assessing the Impacts of New Mobility on Cities.* Report No. NITC-RR-1249. Portland, OR: Transportation Research and Education Center (TREC), 2020. https://www.portlandoregon.gov/transportation/article/709719.

Lyft. "Lyft and Grubhub Team Up to Bring Unlimited Free Delivery From Your Favorite Restaurants and Other New Perks to Lyft Pink." *Blog.* October 6, 2020. https://www.lyft.com/blog/posts/lyftpink-adds-grubhub-perks.

Mass Transit. "SFMTA Launches Pilot Program to Test Adaptive Scooters for People with Disabilities." January 22, 2020. https://www.masstransitmag.com/alt-mobility/shared-mobility/bicycle-scooter-sharing/press-release/21122385.

May, Ethan. "Heres Everything You Need To Know About Bird and Lime Electric Scooters." *IndyStar.* Last modified September 25, 2019. https://www.indystar.com/story/news/2018/06/21/bird-electric-scooters-rental-costs-hours-charging-locations/720893002/.

National Association of City Transportation Officials. *Shared Micromobility in the US: 2019.* New York, NY: National Association of City Transportation Officials (NACTO), 2020. https://nacto.org/wp-content/uploads/2020/08/2020bikesharesnapshot.pdf.

NYC Bike Maps. "Citi Bike." Accessed September 23, 2020. http://www.nycbikemaps.com/citi-bike/.

Upfront Ventures. "Travis VanderZanden Interviewed by Mark Suster | Upfront Summit 2019." March 25, 2019. Video, 25:18. https://www.youtube.com/watch?v=wqWkAXJY_fw.

Zarif, Rasheq, Ben Kelman and Derek Panratz. "Small is Beautiful." Deloitte Insights. Deloitte. April 15, 2019. https://www2.deloitte. com/us/en/insights/focus/future-of-mobility/micro-mobility-is-the-future-of-urban-transportation.html.

CHAPTER 7: MICROTRANSIT

American Public Transportation Association (APTA). "Public Transportation Facts." Accessed October 1, 2020. https://www. apta.com/news-publications/public-transportation-facts/.

American Public Transportation Association (APTA). *Transit Ridership Report: First Quarter 2019.* Washington, DC: APTA, 2019. https://www.apta.com/wp-content/uploads/2019-Q1-Ridership-APTA-1.pdf.

American Public Transportation Association (APTA). *Transit Ridership Report: Fourth Quarter 2019.* Washington, DC: APTA, 2020. https://www.apta.com/wp-content/uploads/2019-Q4-Ridership-APTA.pdf.

American Public Transportation Association (APTA). *Transit Ridership Report: Second Quarter 2019.* Washington, DC: APTA, 2019. https://www.apta.com/wp-content/uploads/2019-Q2-Ridership-APTA.pdf.

American Public Transportation Association (APTA). *Transit Ridership Report: Third Quarter 2019.* Washington, DC: APTA,

2019. https://www.apta.com/wp-content/uploads/2019-Q3-Ridership-APTA.pdf.

Arrillaga, B. and G.E. Mouchahoir, *Demand-Responsive Transportation System Planning Guidelines*. Special Report 136. Washington, DC: National Academy of Sciences - National Research Council, 1974. https://trid.trb.org/view/137385.

INRIX. "INRIX: Congestion Costs Each American 97 hours, $1,348 A Year." INRIX. press release, February 11, 2019. INRIX website. Accessed October 1, 2020. https://inrix.com/press-releases/scorecard-2018-us/.

Koffman, David. *Operational Experiences with Flexible Transit Services*. Washington, DC: Transportation Research Board, 2004. http://onlinepubs.trb.org/onlinepubs/tcrp/tcrp_syn_53.pdf.

Leriche, Yann. "Microtransit: The Next Mobility Revolution or Much Ado About Nothing?" *Medium*. October 12, 2019. https://medium.com/@YannLeriche/microtransit-the-next-mobility-revolution-or-much-ado-about-nothing-d54ad6adf1a.

Marshall, Aarian. "LA Looks to Rideshare to Build the Future of Public Transit." *Wired*. October 24, 2017. https://www.wired.com/story/la-rideshare-public-transit/.

Mass Transit. "COTA, Via partner to provide on-demand service in Central Ohio communities." October 15, 2020. https://www.masstransitmag.com/bus/press-release/21158563/via-transportation-cota-via-partner-to-provide-ondemand-service-in-central-ohio-communities?.

Mibiz. "Pilot Program to Boost Mobility Options for People with Disabilities." July 13, 2020. https://mibiz.com/sections/economic-development/pilot-program-to-boost-mobility-options-for-people-with-disabilities.

Powers, Josh. "Microtransit." September 8, 2019. In *JoCo on the Go: Everything Johnson County Kansas*. Produced by Theresa Freed. Podcast, Podbean, 15:39. https://jocogov.podbean.com/e/microtransit/.

SAE International. "What is Shared Mobility?" Shared Mobility. Accessed October 1, 2020. https://www.sae.org/shared-mobility/.

The Rapid. "New App Offers The Rapids GO!Bus Passengers Convenience and Less Wait Time." The Rapid. press release, August 5, 2019. The Rapid website. Accessed September 8, 2020. https://www.ridetherapid.org/articles/new-app-offers-the-rapids-go-bus-passengers-convenience-and-less-wait-time.

Shuttl. "Shuttl: Stress-Free Commute to Work." Accessed October 5, 2020. https://ride.shuttl.com/.

Web Summit. "Moving Past Private Vehicles and Public Buses." November 6, 2018. Video, 11:32. https://www.youtube.com/watch?v=3EGel2nQNlA.

CHAPTER 8: SHARED MOBILITY AND MOBILITY AS A SERVICE

Boenau, Andy. "Mobility-as-a-Service 101." *Blog*. Accessed October 5, 2020. https://www.andyboenau.com/mobility-as-a-service-101/.

Busvine, Douglas. "From U-Bahn to E-Scooters: Berlin Mobility App Has it All." *Reuters*. September 24, 2019. https://www.reuters.com/article/us-tech-berlin/from-u-bahn-to-e-scooters-berlin-mobility-app-has-it-all-idUSKBN1W90MG.

Essaidi, Mehdi, Claire Duthu, Sebastian Tschödrich, Ross Douglas and Guillaume Cordonnier. "The Future of Mobility as a Service (MaaS): Which model of MaaS Will Win Through?" Capgemini Invent, Capgemini and Autonomy. 2020. https://www.capgemini.com/wp-content/uploads/2020/12/Capgemini-Invent-POV-Maas.pdf.

GlobeNewswire. "Global Mobility As A Service (MaaS) Market 2020-2027 by Service Type, Application, Business Model, Vehicle Type." Research and Markets. December 10, 2020. https://www.globenewswire.com/news-release/2020/12/10/2142838/0/en/Global-Mobility-As-A-Service-MaaS-Market-2020-2027-by-Service-Type-Application-Business-Model-Vehicle-Type.html.

GlobeNewswire. "Global Shared Mobility Market Size & Trends Will Reach to USD 238.03 billion by 2026: Facts & Factors." Facts & Factors. December 10, 2020. https://www.globenewswire.com/news-release/2020/12/10/2143121/0/en/Global-Shared-Mobility-Market-Size-Trends-Will-Reach-to-USD-238-03-billion-by-2026-Facts-Factors.html.

Gudonavičius, Martynas. "Yumuv—the Next Big Leap for Mobility as a Service." Trafi. *Medium.* August 25, 2020. https://medium. com/trafi/yumuv-the-next-big-leap-for-mobility-as-a-service-d4ab97bb6980.

Hartikainen, Ali, Jukka-Pekka Pitkänen, Atte Riihelä, Jukka Räsänen, Ian Sacs, Ari Sirkiä, and Andre Uteng. "Whimpact." *Ramboll.* May 21, 2019. https://ramboll.com/-/media/files/rfi/ publications/Ramboll_whimpact-2019.pdf.

Hook, Leslie. "Expedia Boss Dara Khosrowshahi on the New Breed of Disrupters." *Financial Times.* July 23, 2017. https:// www.ft.com/content/81d319ec-3be4-11e7-821a-6027b8a20f23.

House of Commons of Transport Committee. *Mobility as a Service.* London, U.K.: authority of the House of Commons. 2018. https://publications.parliament.uk/pa/cm201719/cmselect/ cmtrans/590/590.pdf.

Jelbi. "The Jelbi Mobility Partners." Accessed January 5, 2021. https://www.jelbi.de/en/mobility-partners/.

Korosec, Kirsten. "Trafi Takes its Mobility-as-a-Service Platform to LatAm, Starting with Bogota." TechCrunch. February 15, 2021. https://techcrunch.com/2021/02/15/trafi-takes-its-mo-bility-as-a-service-platform-to-latam-starting-with-bogota/.

Kyyti Group. "RTD and Kyyti Group Launch App that Fully Integrates Regular Bus and Rail Services with Flexride Service." Kyyti Group. press release, February 8, 2021. Kyyti Group website. Accessed February 5, 2021. https://www.kyyti.com/

rtd-and-kyyti-group-launch-app-that-fully-integrates-regular-
bus-and-rail-services-with-flexride-service/.

Lango, Luke. "The Era of Car Ownership Is Over. And These 4
Charts Prove It." InvestorPlace. April 3, 2019. https://investor-
place.com/2019/04/4-charts-car-ownership-over/.

MaaS Global Oy. "Find Your Plan." *Whim*. Accessed January 15,
2021. https://whimapp.com/plans/.

MaaS Global Oy. "MaaS Global Completes €29.5M Funding
Round." *Whim (blog)*. November 7, 2019. https://whimapp.
com/maas-global-completes-e29-5m-funding-round/.

Mordor Intelligence. "BFSI Security Market - Growth, Trends,
Covid-19 Impact, and Forecasts (2021–2026)." 2020. https://
www.mordorintelligence.com/industry-reports/bfsi-securi-
ty-market.

SAE International. "What is Shared Mobility?" Shared Mobility.
Accessed October 1, 2020. https://www.sae.org/shared-mobil-
ity/.

Sawers, Paul. "Wunder Mobility Closes $60 Million Round to
Expand its Urban Transport Platform in the US" Venture
Beat. September 19, 2019. https://venturebeat.com/2019/09/19/
wunder-mobility-closes-60-million-round-to-expand-its-ur-
ban-transport-platform-in-the-u-s/.

Sochor, Jana. *Piecing Together the Puzzle Mobility as a Service from
the User and Service Design Perspectives*. International Transport Forum Discussion Papers. No. 2021/08. Paris, France:

OECD Publishing. 2021. https://www.itf-oecd.org/sites/default/files/docs/maas-user-service-design.pdf.

TED. "Ubers Plan to Get More People into Fewer Cars | Travis Kalanick." March 25, 2016. Video, 19:18. https://www.youtube.com/watch?v=pb--rJGgVIo.

United Nations. "2018 Revision of World Urbanization Prospects." Department of Economic and Social Affairs. May 16, 2018. https://www.un.org/development/desa/publications/2018-revision-of-world-urbanization-prospects.html.

United States Department of Transportation Federal Highway Administration. "State Motor-Vehicle Registrations 2018." Policy and Governmental Affairs Office of Highway Policy Information. December, 2019. https://www.fhwa.dot.gov/policyinformation/statistics/2018/pdf/mv1.pdf.

Valdes, Vincent. "The Impact of Innovation on MaaS/MOD with Vincent Valdes." June 23, 2020. In *ITE Talks Transportations Tracks.* Produced by Bernie Wagenblast. Podcast, Spreaker, 19:05. https://www.spreaker.com/user/ite-talks-transportation/valdes-june.

CHAPTER 9: WHY MOBILITY AS A SERVICE?

American Automobile Association (AAA). "Your Driving Costs, 2020." December 14, 2020. https://newsroom.aaa.com/wp-content/uploads/2020/12/Your-Driving-Costs-2020-Fact-Sheet-FINAL-12-9-20-2.pdf.

Böhm, Steffen, Campbell Jones and Christopher Land. "Part One Conceptualizing Automobility: Introduction: Impossibilities of Automobility." *Sociological Review* (September 2006). https://doi.org/10.1111/j.1467-954X.2006.00634.x.

Buchholz, Katharina. "Americans Get Driver's licenses Later in Life." Statista. January 7, 2020. https://www.statista.com/chart/18682/percentage-of-the-us-population-holding-a-drivers-license-by-age-group/.

Carlton Reid. "Netflix-Of-Transportation App Reduces Car Use And Boosts Bike And Bus Use, Finds MaaS Data Crunch." *Forbes*. March 28, 2019. https://www.forbes.com/sites/carltonreid/2019/03/28/netflix-of-transportation-app-reduces-car-use-and-boosts-bike-and-bus-use-finds-maas-data-crunch/.

Expat Finland. "Driving in Finland: Licences, Rules, Vehicles & Tyres, Schools." Accessed February 10, 2021. https://www.expat-finland.com/living_in_finland/driving.html#.

Futurebuilders Podcast. "Interview with Sampo Hietanen—Mobility as a Service, the Future of Transportation." April 23, 2019. Video, 38:14. https://www.youtube.com/watch?v=voOuBrFhNco.

Hartikainen, Ali, Jukka-Pekka Pitkänen, Atte Riihelä, Jukka Räsänen, Ian Sacs, Ari Sirkiä, and Andre Uteng. "Whimpact." *Ramboll*. May 21, 2019. https://ramboll.com/-/media/files/rfi/publications/Ramboll_whimpact-2019.pdf.

Henderson, Tim. "Why Many Teens Don't Want to Get a Driver's license." *PBS*. March 6, 2017. https://www.pbs.org/newshour/nation/many-teens-dont-want-get-drivers-license.

Hietanen, Sampo. Quoted in Carlton Reid. "Netflix-Of-Transportation App Reduces Car Use And Boosts Bike And Bus Use, Finds MaaS Data Crunch." *Forbes*. March 28, 2019. https://www.forbes.com/sites/carltonreid/2019/03/28/netflix-of-transportation-app-reduces-car-use-and-boosts-bike-and-bus-use-finds-maas-data-crunch/.

MaaS Global Oy. "A Family of Five Gave Up their Car: Life Now Runs Smoothly for Them with Good Cycling Equipment and the Occasional Rental Car." *Whim (blog)*. September 21, 2020. https://whimapp.com/a-family-of-five-gave-up-their-car-life-now-runs-smoothly-for-them-with-good-cycling-equipment-and-the-occasional-rental-car/.

MaaS Global Oy. "CEO Sampo Hietanen Honoured with Finland's Order of the White Rose." *Whim (blog)*. January 14, 2020. https://whimapp.com/ceo-sampo-hietanen-honoured-with-finlands-order-of-the-white-rose/.

MaaS Global Oy. "Easy Access by Public Transport and Bike—Young People are Postponing Obtaining a Driving License." *Whim (blog)*. November 7, 2019. https://whimapp.com/easy-access-by-public-transport-and-bike-young-people-are-postponing-obtaining-a-driving-licence/.

Sachs, Wolfgang. *For Love of the Automobile: Looking Back into the History of our Desires*. Translated by Don Reneau. Los Angeles: University of California Press, 1992. https://books.

google.com/books?hl=en&lr=&id=gGx2wcTxj9UC&oi=f-nd&pg=PR7&ots=mE8rV61beZ&sig=SMpQdoKSV2NljCL-4VI1TK8yfOkQ#v=onepage&q&f=false.

TheCamp. "Disruption in Mobility and Why it Will Be Collaborative." October 30, 2020. Video, 34:38. https://vimeo.com/473761053.

Urry, John. *Mobilities* (Cambridge, USA: Polity Press, 2008). Quoted in Audouin, Maxime and Matthias Finger. "The Development of Mobility-as-a-Service in the Helsinki Metropolitan Area: A Multi-Level Governance Analysis." Research in Transportation Business & Management. Elsevier, 2018. https://doi.org/10.1016/j.rtbm.2018.09.001.

Urry, John. "The System of Automobility." *Theory Culture & Society* (October 2004). https://doi.org/10.1177/0263276404046059.

Valdes, Vincent. "The Impact of Innovation on MaaS/MOD with Vincent Valdes." June 23, 2020. In *ITE Talks Transportations Tracks.* Produced by Bernie Wagenblast. Podcast, Spreaker, 19:05. https://www.spreaker.com/user/ite-talks-transportation/valdes-june.

CHAPTER 10: REALIZING MAAS

American Public Transportation Association (APTA). "Being Mobility-as-a-Service (MaaS) Ready." Washington, DC: American Public Transportation, 2019. https://www.apta.com/wp-content/uploads/MaaS_European_Study_Mission-Final-Report_10-2019.pdf.

Autonomy. "How Europe is Moving MaaS -and Vice Versa in partnership with Hogan Lovells." October 27, 2020. Video, 1:03:46. https://www.youtube.com/watch?v=1_EzTal77h4.

Bach, David and Bruce Allen. "What Every CEO Needs to Know About Nonmarket Strategy." *Research Feature.* MIT Sloan Management Review. April 1, 2010. https://sloanreview.mit. edu/article/what-every-ceo-needs-to-know-about-nonmarket-strategy/.

Boenau, Andy. "Why can't Anyone Build a Thriving and Sustainable MaaS Business?" Urban Mobility. January 13, 2021. https:// urbanmobilitycompany.com/content/daily/why-cant-anyone-build-a-thriving-and-sustainable-maas-business.

Cao, Sissi. "Elon Musk's Tunnel Under Las Vegas for Self-Driving Cars Is Almost Complete." *Observer.* September 16, 2020. https://observer.com/2020/09/elon-musk-boring-company-tunnel-las-vegas-near-completion/.

City Tech Collaborative. "City Tech Launches New Solution to Address Urban Curbside Chaos." January 29, 2020. https:// www.citytech.org/city-tech-launches-new-solution-to-address-urban-curbside-chaos.

"Final Dockless Scooter Terms and Conditions." Government of the District of Columbia Department of Transportation. 2021. https://ddot.dc.gov/sites/default/files/dc/sites/ddot/page_content/attachments/Final%20Dockless%20Scooter%20Terms%20and%20Conditions.pdf.

Institute of Transportation Engineers (ITE). "Curbside Management Resources." Accessed February 21, 2021. https://www. ite.org/technical-resources/topics/complete-streets/curbside-management-resources/#.

Kamargianni, Maria and Melinda Matyas. *The Business Ecosystem of Mobility as a Service.* Washington, DC: *Transportation Research Board (TRB),* 2017. https://www.researchgate.net/ publication/314760234_The_Business_Ecosystem_of_Mobility-as-a-Service.

Klender, Joey. "The Boring Co.'s Projects Are Making Transit Departments Rethink Above-Ground Travel." Teslarati. February 3, 2021. https://www.teslarati.com/boring-company-elon-musk-above-ground-passenger-transport-rethink/.

MaaS Alliance. "The Alliance." Accessed October 8, 2020. https:// maas-alliance.eu/the-alliance/.

MaaS Alliance. "Working Together." Accessed October 8, 2020. https://maas-alliance.eu/homepage/working-together/.

MaaS America. "MaaS America: Advancing the New Mindset of Mobility in America." Accessed October 8, 2020. https://www. maasamerica.org/.

MaaS America. "What We Do." Accessed October 8, 2020. https:// www.maasamerica.org/whatwedo.

McGuckin, Tim. "Mobility-as-a-Service An Interview with MaaS Americas Tim McGuckin." October 28, 2019. In *The Smart City Podcast.* Produced by Eduard Fidler. Podcast, *ARC Advisory*

Group, 1:18:33. https://www.arcweb.com/blog/mobility-service-interview-maas-americas-tim-mcguckin.

Merano, Maria. "The Boring Company Talks for Ontario Loop Project Begins in San Bernardino County." Teslarati. February 4, 2021. https://www.teslarati.com/the-boring-company-ontario-loop-in-san-bernardino-county/.

Mulley, Corinne and John Nelson, *How Mobility as a Service Impacts Public Transport Business Models.* International Transport Forum Discussion Papers. No. 2020/17. Paris, France: OECD Publishing, 2020. https://www.itf-oecd.org/sites/default/files/docs/maas-impacts-public-transport-business-models.pdf.

National Association of City Transportation Officials (NACTO). "Curb Appeal: Curbside Management Strategies for Improving Transit Reliability." Accessed February 21, 2021. https://nacto.org/tsdg/curb-appeal-whitepaper/.

Pew Research Center. "Mobile Fact Sheet." June 12, 2019. https://www.pewresearch.org/internet/fact-sheet/mobile/#.

Sheetz, Michael. "NASA Estimates Having SpaceX and Boeing Build Spacecraft for Astronauts Saved $20 Billion to $30 Billion." *CNBC.* Last modified May 13, 2020. https://www.cnbc.com/2020/05/13/nasa-estimates-having-spacex-and-boeing-build-spacecraft-for-astronauts-saved-up-to-30-billion.html#.

Sheetz, Michael. "SpaceX is about to Launch its First Full NASA Crew to the Space Station: Here's What You Should Know." *CNBC.* Last modified November 15, 2020. https://www.cnbc.

com/2020/11/15/spacex-and-nasa-crew-1-launch-heres-what-you-should-know.html.

"Infrastructure Prioritization." StreetLight Data. Accessed February 25, 2021. https://www.streetlightdata.com/transportation-planning-traffic-planners-product/#infrastructure-prioritization.

TheCamp. "Disruption in Mobility and Why it Will Be Collaborative." October 30, 2020. Video, 34:38. https://vimeo.com/473761053.

The National Academies of Sciences, Engineering, and Medicine. "About | Transportation Research Board." Accessed October 8, 2020. https://www.nationalacademies.org/trb/about.

Smith, G. *Making Mobility-as-a-Service: Towards Governance Principles and Pathways.* Doctoral Thesis. Gothenburg, Sweden: Chalmers University of Technology, 2020. Quoted in Corinne Mulley et al. *How Mobility as a Service Impacts Public Transport Business Models.* International Transport Forum Discussion Papers. No. 2020/17. Paris, France: OECD Publishing, 2020.

CHAPTER 11: ELECTRIFYING AND AUTOMATING MAAS

Bay, Oyster Bay. "ABI Research Forecasts Global Mobility as a Service Revenues to Exceed $1 Trillion by 2030." ABI Research, September 12, 2016. https://www.abiresearch.com/press/abi-research-forecasts-global-mobility-service-rev/.

Churchill, Winston Churchill. Quoted in "History Repeating." College of Liberal Arts and Human Sciences. Virginia Tech.

Accessed January 15, 2021. https://liberalarts.vt.edu/magazine/2017/history-repeating.html.

Corwin, Scott, Nick Jameson, Derek Pankratz and Philipp Willigmann. "The Future of Mobility: What's Next?" Deloitte Insights. Deloitte. September 14, 2016. https://www2.deloitte.com/us/en/insights/focus/future-of-mobility/roadmap-for-future-of-urban-mobility.html.

Essaidi, Mehdi, Claire Duthu, Sebastian Tschödrich, Ross Douglas and Guillaume Cordonnier. "The Future of Mobility as a Service (MaaS): Which model of MaaS Will Win Through?" Capgemini Invent, Capgemini and Autonomy. 2020. https://www.capgemini.com/wp-content/uploads/2020/12/Capgemini-Invent-POV-Maas.pdf.

Futurebuilders Podcast. "Interview with Sampo Hietanen—Mobility as a Service, the Future of Transportation." April 23, 2019. Video, 38:14. https://www.youtube.com/watch?v=voOuBrFhNco.

Hannon, Eric, Stefan Knupfer, Sebastian Stern and Jan Tijs Nijssen. "The Road to Seamless Urban Mobility." McKinsey & Company. January 16, 2019. https://www.mckinsey.com/business-functions/sustainability/our-insights/the-road-to-seamless-urban-mobility.

Litman, Todd. *Autonomous Vehicle Implementation Predictions: Implications for Transport Planning.* Victoria, Canada: Victoria Transport Policy Institute, 2020. https://www.vtpi.org/avip.pdf.

TheCamp. "Disruption in Mobility and Why it Will Be Collaborative." October 30, 2020. Video, 34:38. https://vimeo.com/473761053.

Toyota. "Toyota Shows e-Palette Geared Towards Practical MaaS Application." Toyota. press release. December 22, 2020. Toyota website. Accessed January 8, 2021. https://global.toyota/en/newsroom/corporate/34527341.html.

Toyota. "Toyota Woven City." Accessed January 8, 2021. https://www.woven-city.global/.

CONCLUSION

Chase, Robin. Quoted in "Reimagining Public Transit.", Mission. *Medium.* December 14, 2018. https://medium.com/the-mission/reimagining-public-transit-a5ac5e5bd440.